Molecular Methods of Plant Analysis

Editors:
J.F. Jackson (Managing Editor)
H.F. Linskens
R.B. Inman

Volume 23

Springer

Berlin
Heidelberg
New York
Hong Kong
London
Milan
Paris
Tokyo

Volumes Already Published in this Series (formerly Modern Methods of Plant Analysis):

Volume 1: Cell Components
1985, ISBN 3-540-15822-7
Volume 2: Nuclear Magnetic Resonance
1986, ISBN 3-540-15910-X
Volume 3: Gas Chromatography/Mass Spectrometry
1986, ISBN 3-540-15911-8
Volume 4: Immunology in Plant Sciences
1986, ISBN 3-540-16842-7
Volume 5: High Performance Liquid Chromatography in Plant Sciences
1986, ISBN 3-540-17243-2
Volume 6: Wine Analysis
1988, ISBN 3-540-18819-3
Volume 7: Beer Analysis
1988, ISBN 3-540-18308-6
Volume 8: Analysis of Nonalcoholic Beverages
1988, ISBN 3-540-18820-7
Volume 9: Gases in Plant and Microbial Cells
1989, ISBN 3-540-18821-5
Volume 10: Plant Fibers
1989, ISBN 3-540-18822-3
Volume 11: Physical Methods in Plant Sciences
1990, ISBN 3-540-50332-3
Volume 12: Essential Oils and Waxes
1991, ISBN 3-540-51915-7
Volume 13: Plant Toxin Analysis
1992, ISBN 3-540-52328-6
Volume 14: Seed Analysis
1992, ISBN 3-540-52737-0
Volume 15: Alkaloids
1994, ISBN 3-540-52738-9
Volume 16: Vegetables and Vegetable Products
1994, ISBN 3-540-55843-8
Volume 17: Plant Cell Wall Analysis
1996, ISBN 3-540-59406-X
Volume 18: Fruit Analysis
1995, ISBN 3-540-59118-4
Volume 19: Plant Volatile Analysis
1997, ISBN 3-540-61589-X
Volume 20: Analysis of Plant Waste Materials
1999, ISBN 3-540-64669-8
Volume 21: Analysis of Taste and Aroma
2002, ISBN 3-540-41753-2
Volume 22: Testing for Genetic Manipulation in Plants
2002, ISBN 3-540-43153-5
Volume 23: Genetic Transformation of Plants
2003, ISBN 3-540-00292-8

Genetic Transformation of Plants

Edited by
J.F. Jackson and H.F. Linskens

With 21 Figures, 4 in Color
and 11 Tables

Professor J.F. Jackson
Department of Horticulture
Viticulture and Oenology
University of Adelaide
Waite Campus
SA 5064 Glen Osmond
Australia

Professor H.F. Linskens
Goldberglein 7
91056 Erlangen
Germany

Professor R.B. Inman
Institute of Molecular Virology
University of Wisconsin-Madison
Robert M. Bock Laboratories
1525 Linden Drive
Madison, Wisconsin 53706-1596
USA

ISSN 1619-5221
ISBN 3-540-00292-8 Springer-Verlag Berlin Heidelberg New York

Cataloging-in-Publication Data applied for

A catalog record for this book is available from Library of Congress.
Bibliographic information published by Die Deutsche Bibliothek.
Die Deutsche Bibliothek lists this publication in the Deutsche Nationalbibliografie; detailed bibliographic data is available in the Internet at http://dnb.ddb.de

Library of Congress Cataloging-in-Publication Data

Genetic transformation of plants / edited by J.F. Jackson and H.F. Linskens.
p. cm. – (Molecular methods of plant analysis, ISSN 1619-5221 ; v. 23)
Includes bibliographical references and index.
ISBN 3-540-00292-8 (alk. paper)
1. Plant genetic transformation. 2. Crops–Genetic engineering. 3. Plant genetic engineering. I. Jackson, J. F. (John F.), 1935- II. Linskens, H. F. (Hans F.), 1921- III. Series.

QK865.M57 vol. 23
[SB123.57]
571.2′028 s–dc21
[631.5′233]

2003042766

Springer-Verlag Berlin Heidelberg New York
a member of BertelsmannSpringer Science+Business Media GmbH
http://www.springer.de

Printed in Germany

Production: PRO EDIT GmbH, Heidelberg

Cover design: design & production GmbH, Heidelberg
Cover photograph: Dr. Malin Elfstrand and Mr Hartmut Weichelt
Typesetting: SNP Best-set Typesetter Ltd., Hong Kong
Printed on acid-free paper 31/3150 Di – 5 4 3 2 1 0

Preface

Molecular Methods of Plant Analysis

Concept of the Series

The powerful recombinant DNA technology and related developments have had an enormous impact on molecular biology. Any treatment of plant analysis must make use of these new methods. Developments have been so fast and the methods so powerful that the editors of *Modern Methods of Plant Analysis* have now decided to rename the series *Molecular Methods of Plant Analysis*. This will not change the general aims of the series, but best describes the thrust and content of the series as we go forward into the new millennium. This does not mean that all chapters a priori deal only with the methods of molecular biology, but rather that these methods are to be found in many chapters together with the more traditional methods of analysis which have seen recent advances. The numbering of the volumes of the series therefore continues on from 20, which is the most recently published volume under the title *Modern Methods of Plant Analysis*.

As indicated for previous volumes, the methods to be found in *Molecular Methods of Plant Analysis* are described critically, with hints as to their limitations, references to original papers and authors being given, and the chapters written so that there is little need to consult other texts to carry out the methods of analysis described. All authors have been chosen because of their special experience in handling plant material and/or their expertise with the methods described. The volumes of the series published up to now fall into three groups: Volumes 1–5 and Volume 11 dealing with some basic principles of methods, Volumes 6, 7, 8, 10, 14, 16, 18 and 20 being a group determined by the raw plant material being analysed, and a third group comprising Volumes 9, 12, 13, 15, 17 and 19 which are separated from the other volumes in that the class of substances being analysed, is indicated in the volume title. Volume 21 and future volumes of *Molecular Methods of Plant Analysis* will continue in a similar vein but will include more chapters involved with the methods of molecular biology.

Development of the Series

The handbook, *Modern Methods of Plant Analysis*, was first introduced in 1954, and was immediately successful, seven volumes appearing between 1956 and 1964. This first series was initiated by Michael Tracey of Rothamsted and Karl Paech of Tübingen. The so-called *New Series of Modern Methods of Plant Analysis*, Volumes 1–20, began in 1985 and has been edited by Paech's successor, H.F. Linskens of Nijmegen, The Netherlands, and John F. Jackson of Adelaide, South Australia. These same editors have now teamed up with a third, Ross B. Inman of Madison, Wisconsin, USA, to produce the renamed series *Molecular Methods of Plant Analysis*. As before, the editors are convinced that there is a real need for a collection of reliable, up-to-date methods of plant analysis covering large areas of applied biology ranging from agricultural and horticultural enterprises to pharmaceutical and technical organizations concerned with material of plant origin.

Future volumes will include Various Aspects of Plant Genomics.

Volume 23: *Genetic Transformation of Plants*

This third volume in the molecular series deals with the topic of genetic transformation of plants. Most would view genetic transformation as a means of bringing about plant improvement, however, it can be a useful tool in analysing the function of plant genes. To this end, the present volume focuses on genetic transformation of a range of plants by a range of methods, a multiplicity of methods being necessary as some plants are more difficult to transform than others.

Since in genetic transformation we are dealing with biotechnological innovation, this volume begins with a chapter on "Biotechnology, Genetic Manipulation and Intellectual Property Rights". It is beyond dispute that property rights apply to the products of biological research, and there is no doubt that in the "developed" world DNA sequences and cells of plant or animal origin can be patented. This first chapter then explores these property rights, be they physical or intellectual, and how they effect the use to which the transformed plant is put and the right to reproduce it.

The following chapter describes the many methods used to carry out plant transformation, beginning with *Agrobacterium rhizogenes*-mediated transformation, which leads to "hairy root" syndrome. This is particularly useful in analysing the interaction between roots and soil organisms or chemical compounds. Thus, promoter-trapping strategies using hairy roots have been utilized to identify genes that form nodules, while hairy roots have also been used to study the interaction between roots and nematodes. Analysis of responses to such chemicals as fungicides, nematicides and herbicides can also utilize the hairy root condition. The next chapter deals with *Agrobacterium tumefaciens*-mediated transformation of whole plants of *Petunia hybrida*, in this case, by a suspension of *Agrobacterium* cells applied directly to the flower stigma at pollination. *Allium* species have well developed sulphur and carbohydrate biochemical pathways which need to be thoroughly investigated. However, *Allium* has proven to be very difficult to transform; a chapter on *Allium* transformation is therefore included in the belief that it will assist in the analysis of these pathways and identify which are important for normal physiology and which are crucial for the unique neutraceutical qualities ascribed to garlic and onions. Similarly, barley has proven difficult to transform, thus an electroporation method is described in this volume for barley.

Sorghum, like barley, proved difficult to transform at first. A chapter therefore follows on transformation of sorghum using *Agrobacterium tumefaciens.* Polyethylene glycol (PEG) was amongst the first gene transfer systems to be used for successful integration of foreign genes into plant cells. A chapter is included in this volume describing effective production of transgenic sunflower by a PEG-mediated transformation system. However, a large number of sunflower protoplasts need to be used to ensure a significant number of transformed plants. Sunflower exhibits considerable sexual incompatibility between crop and wild species, which limits access to the genetic pool for gene analysis (or plant improvement), and so transformation provides an alternative approach. The last few chapters of this book deal with particle bombardment and WHISKERS-mediated methods of transformation. Norway spruce transformation can be carried out by particle bombardment of embryonic cultures or pollen; the method is important in developing better or new qualities of wood. The author also discusses the considerable problems associated with gene flow by pollen following spruce transformation. A chapter follows on WHISKERS transformation of embryonic maize suspension cultures leading to regeneration into fertile transgenic plants. A subsequent chapter deals with genetic transformation of soybean with biolistics. The latter involves bombardment of proliferative embryonic cultures with DNA coated on 1-μm diameter particles followed by selection and plant regeneration. Both spruce and soybean biolistic transformation methods described above utilized gold particles coated with DNA, although tungsten particles were used in the past. It seems that tungsten, unlike gold, causes considerable DNA damage including DNA strand scissions and inhibition of cell differentiation; these and other genotoxic effects of tungsten particles are assessed in the final chapter.

J.F. Jackson, H.F. Linskens, R.B. Inman

Contents

1 Exclusive Rights in Life: Biotechnology, Genetic Manipulation, and Intellectual Property Rights
E.R. GOLD

1.1 Introduction ... 1
1.2 Biotechnological Innovation ... 2
1.2.1 Physical Innovations ... 3
1.2.1.1 DNA and Protein Molecules ... 3
1.2.1.2 Cells ... 3
1.2.1.3 Whole Organisms ... 4
1.2.2 Information and Other Intangibles ... 4
1.2.2.1 DNA Sequences and Cells ... 4
1.2.2.2 Processes Using Biological Matter ... 5
1.2.2.3 Bioinformatics ... 5
1.2.3 Summary ... 6
1.3 Introduction to Intellectual Property Rights ... 6
1.3.1 Exclusive Rights vs. Rights to Things ... 6
1.3.2 Property and Intellectual Property Rights ... 7
1.3.3 Trade Secrets ... 7
1.3.3.1 Subject Matter ... 8
1.3.3.2 Requirements ... 8
1.3.4 Patents ... 8
1.3.4.1 Subject Matter ... 9
1.3.4.1.1 Invention vs. Discovery ... 10
1.3.4.1.2 Exclusions ... 10
1.3.4.2 Requirements ... 11
1.3.4.2.1 Substantive Criteria ... 11
1.3.4.2.1.1 Novelty ... 12
1.3.4.2.1.2 Inventive Step (Nonobviousness) ... 12
1.3.4.2.1.3 Industrial Application (Utility) ... 12
1.3.4.2.2 Procedural Criterion: Disclosure ... 13
1.3.4.3 Remedies ... 14
1.3.5 Copyright and Database Protection ... 14
1.3.5.1 Subject Matter ... 15

1.3.5.2 Requirements 15
1.3.5.3 Remedies 16
1.3.6 Plant Variety Protection 16
1.3.6.1 Subject Matter 16
1.3.6.2 Requirements 17
1.4 Challenges 17
1.4.1 Incentive vs. Access 18
1.4.1.1 Justification for Property Rights 18
1.4.1.2 Economic Reality 19
1.4.2 Fairness to Providers of Biological Matter 20
1.4.2.1 Rights to Biological Matter 20
1.5 Conclusion 21
References 21

2 *Agrobacterium rhizogenes*-Mediated Transformation of Plants
W. VAN DE VELDE, M. KARIMI, G. DEN HERDER, M. VAN MONTAGU,
M. HOLSTERS, and S. GOORMACHTIG

2.1 Introduction 23
2.2 Aspects Influencing *A. rhizogenes* Transformation Efficiency 26
2.2.1 Choice of *A. rhizogenes* Strain 26
2.2.2 Choice of Explant 28
2.2.3 Preparation of Bacterial Inoculum and Infection of Explants 29
2.2.4 Co-cultivation 29
2.3 Establishing the Transformed Nature of Hairy Roots 31
2.4 Cotransformation of Binary T-DNA 31
2.5 Propagation of Hairy Root Lines in Liquid Cultures 33
2.5.1 The Clonal Status of Hairy Roots 33
2.5.2 Stability of Long-Term Hairy Root Cultures 34
2.6 Regeneration of Plants from Hairy Roots 34
2.7 The Multi-Auto-Transformation (MAT) Vector System 35
2.8 Conclusions 37
Protocol 1: Production of Transformed Hairy Roots 37
Protocol 2: Plant Regeneration from Hairy Roots 38
Protocol 3: Hairy Root Liquid Culture 39
References 39

3 Transformation of *Petunia hybrida* by the *Agrobacterium* Suspension Drop Method
S.J. WYLIE, D. TJOKROKUSUMO, and J.A. MCCOMB

3.1 Introduction 45
3.2 Transformation 47
3.3 Analysis of Transformants 47

3.3.1 Screening *Petunia* Seedlings for Herbicide Resistance 47
3.3.2 Transmission of Basta Resistance Phenotype to T_2 Progeny 48
3.3.3 β-Glucuronidase Assay 48
3.3.4 DNA Analysis 49
3.4 Conclusion 49
References 49

4 Onion, Leek and Garlic Transformation by Co-cultivation with *Agrobacterium*
C.C. Eady

4.1 Introduction 53
4.1.1 Current Applications of *Allium* Transformation Technology 53
4.1.1.1 Physiological Studies 53
4.1.1.2 Herbicide Resistance 54
4.1.1.3 Antimicrobial Resistance 54
4.1.1.4 Insect Resistance 55
4.2 Onion Transformation Protocols 55
4.2.1 Transformation Using Antibiotic and Visual Selection 56
4.2.1.1 Bacterial Strain and Plasmids 56
4.2.1.2 Transformation Procedure (Modified from Eady et al. 2000 56
4.2.2 Transformation Using Herbicide Selection 57
4.2.2.1 Bacterial Strain and Plasmids 58
4.2.2.2 Transformation Procedure 58
4.2.3 Ex-Flasking and Growth in Containment 58
4.2.4 Transgene Detection 59
4.2.5 Transgene Expression and Stability 60
4.2.5.1 Visual Reporter Genes 60
4.2.5.2 Expression of Herbicide Resistance 61
4.2.5.3 Antisense Alliinase Gene Expression 61
4.3 Leek Transformation 62
4.4 Garlic Transformation Protocol 63
4.4.1 Bacterial Strain and Plasmids 63
4.4.2 Transformation Procedure 63
4.5 Concluding Remarks 64
References 65

5 Electroporation Transformation of Barley
F. Gürel and N. Gözükirmizi

5.1 Introduction 69
5.2 Background of Electroporation Procedures 71

5.2.1 Pre- and Post-Electroporation Period 71
5.2.2 Electrical Variables 73
5.3 Culture and Electroporation of Barley Explants 74
5.3.1 Protoplasts 74
5.3.2 Microspores 76
5.3.3 Intact Tissues 77
5.3.3.1 Analysis and Inheritance of Transgenes in Electroporated Tissues 81
5.4 Conclusions and Future Perspectives 83
References 84

6 Sorghum Transformation
Z. Zhao and D. Tomes

6.1 Introduction 91
6.2 Sorghum Transformation Process and Optimization 92
6.2.1 Plant Materials and Transformation Systems 92
6.2.2 Transformation Via Microprojectile Bombardment 93
6.2.3 *Agrobacterium*-Mediated Transformation 94
6.3 Analysis of Transgenic Plants and the Progeny 96
6.3.1 Molecular Analysis of T_0 Plants 97
6.3.2 Foreign Gene Expression in T_0 Plants 98
6.3.3 Genetic and Molecular Analysis of the Progeny 99
6.4 Marker-Free Sorghum Transgenic Plants 99
6.4.1 Importance of Marker-Free Transgenics in Sorghum 100
6.4.2 Methods to Eliminate Markers from Transgenic Plants 100
6.4.3 *Agrobacterium* 2 T-DNA Co-Transformation System 101
References 102

7 Transgenic Sunflower: PEG-Mediated Gene Transfer
P.C. Binsfeld

7.1 Introduction 109
7.2 Genetic Variability and Transgenic Breeding 109
7.3 Gene Transfer Systems 110
7.3.1 PEG-Mediated Gene Transfer 111
7.3.1.1 Short DNA Molecule Uptake 111
7.3.1.2 Large DNA Molecule Uptake 112
7.4 Plant Regeneration 114
7.5 General Analytical Considerations 115
7.5.1 Molecular Analysis 115
7.5.1.1 DNA Extraction 116
7.5.1.2 Southern Hybridization 116
7.5.1.3 Polymerase Chain Reaction 116
7.5.1.4 Random Amplified Polymorphic DNA 117

7.5.2 Biochemical Analysis ... 118
7.5.2.1 Multiple Molecular Forms of Enzymes ... 118
7.5.2.2 Enzymatic Assay ... 119
7.5.3 Cytogenetic Analysis ... 120
7.5.3.1 Flow Cytometric Analysis ... 120
7.5.3.2 Mitotic and Meiotic Cell Analysis ... 122
7.5.3.3 In Situ Hybridization ... 122
7.5.4 Morphological Analysis ... 123
7.6 Conclusions and Future Perspectives ... 124
References ... 124

8 Transformation of Norway Spruce (*Picea abies*) by Particle Bombardment
D.H. Clapham, H. Häggman, M. Elfstrand, T. Aronen, and S. von Arnold

8.1 Introduction ... 127
8.2 Types of Particle Accelerator ... 127
8.3 Transformation of Embryogenic Cultures ... 128
8.3.1 Transient Expression in Embryogenic Cultures ... 128
8.3.2 Production of Stably Transformed Cell Cultures and Transgenic Plants ... 129
8.3.3 Stability of Transgene Expression ... 131
8.3.4 Trends in Transgenic Plant Production ... 131
8.4 Transformation of Pollen ... 133
8.4.1 The Reproductive Biology of Norway Spruce ... 133
8.4.2 Transient Expression in Pollen ... 134
8.4.3 Development of Controlled Pollination Techniques for Bombarded Pollen ... 135
8.5 Applications of Transgenic Norway Spruce in Research ... 136
8.5.1 Genes Regulating Embryogenesis ... 136
8.5.2 Genes with Similarity to Defense Genes ... 137
8.6 Prospects for Transgenic Norway Spruce in Practical Forestry ... 139
References ... 143

9 WHISKERS-Mediated Transformation of Maize
J.F. Petolino, M. Welter, and C. Qihua Cai

9.1 Introduction ... 147
9.2 Preparation of Purified DNA Fragments ... 147
9.3 Establishment and Maintenance of Embryogenic Suspension Cultures ... 150
9.4 DNA Delivery via WHISKERS ... 152
9.5 Transgene Copy Number Estimation ... 153
9.6 Regeneration of Transgenic Plants and Progeny ... 157

9.7 Conclusions and Future Perspectives 157
References 158

10 Genetic Transformation of Soybean with Biolistics
D. SIMMONDS

10.1 Introduction 159
10.2 Tissue Culture and Plant Regeneration 160
10.2.1 Genotype Specificity 160
10.2.2 Initiation and Repetitive Proliferation of Somatic Embryogenic Cultures 161
10.2.3 Embryo Histodifferentiation and Maturation 163
10.2.4 Germination, Conversion and Plant Fertility 163
10.3 Transformation 164
10.3.1 Gene Delivery 164
10.3.2 Target Tissue Optimization 165
10.3.3 Selection 165
10.3.4 Transgenic Plant Recovery 166
10.4 Conclusions 167
10.5 Protocol 168
10.5.1 Induction and Maintenance of Proliferative Embryogenic Cultures 168
10.5.2 Transformation 168
10.5.3 Selection 169
10.5.4 Plant Regeneration 169
References 170

11 Genotoxic Effects of Tungsten Microparticles Under Conditions of Biolistic Transformation
J. BUCHOWICZ and C. KRYSIAK

11.1 Introduction 175
11.2 Biological Significance of Tungsten 175
11.2.1 Early Observations on Biological Effects of Tungsten 175
11.2.2 Catalytic Activity of Simple Tungsten Compounds 176
11.2.3 Tungstoenzymes 176
11.2.4 Tungsten–DNA Interaction 177
11.3 Tungsten Microparticles in Biotechnological Applications 178
11.3.1 Biolistic Transformation 178
11.3.1.1 An Overview 178
11.3.1.2 Technical Details 180
11.3.2 Biolistic Inoculation and Related Applications of Tungsten Particles 181
11.4 Assessment of Tungsten-Induced DNA Lesions 182
11.4.1 Electrophoretic Analysis of Tungsten-Damaged Plasmid DNA 182

11.4.2 A Modified TUNEL Method for Detection of Cellular DNA Fragmentation 184
11.5 Post-Bombardment Inhibition of Somatic Embryogenesis 186
11.6 Concluding Remarks 188
References 188

Subject Index 195

List of Contributors

T. Aronen
Finnish Forest Research Institute, Punkaharju Research Station, Finlandiantie 18, 58450 Punkaharju, Finland

P.C. Binsfeld
Center of Biotechnology, Federal University of Pelotas – UFPel, Brazil, Campus Universitário, Caixa Postal 354, 96010-900, Pelotas-RS, Brazil

J. Buchowiczs
Institute of Biochemistry and Biophysics, Polish Academy of Sciences, 5A Pawinskiego Street, 02-106 Warsaw, Poland

D.H. Clapham
Department of Plant Biology and Forest Genetics, Swedish University of Agricultural Sciences, Box 7080, 750 07 Uppsala, Sweden

G. den Herder
Department of Plant Systems Biology, Flanders Interuniversity Institute for Biotechnology, Ghent University, K.L. Ledeganckstraat 35, 9000 Ghent, Belgium

C.C. Eady
New Zealand Institute for Crop & Food Research Limited, Private Bag 4704, Christchurch, New Zealand

M. Elfstrand
Department of Plant Biology and Forest Genetics, Swedish University of Agricultural Sciences, Box 7080, 750 07 Uppsala, Sweden

N. Gözükirmizi
TUBITAK, Research Institute for Genetic Engineering and Biotechnology 2141470 Gebze, Kocaeli, Turkey

E.R. Gold
Bell Chair in e-Governance, Faculty of Law, McGill University, 3664 Peel Street, Montreal, H3A 1W9, Canada

S. GOORMACHTIG
Department of Plant Systems Biology, Flanders Interuniversity Institute for Biotechnology, Ghent University, K.L. Ledeganckstraat 35, 9000 Ghent, Belgium

F. GÜREL
Istanbul University, Department of Molecular Biology and Genetics 34459 Vezneciler, Istanbul, Turkey

H. HÄGGMAN
Finnish Forest Research Institute, Punkaharju Research Station, Finlandiantie 18, 58450 Punkaharju, Finland and Department of Biology, University of Onlu, P. O. Box 3000, 90014 Onlu, Finland

M. HOLSTERS
Department of Plant Systems Biology, Flanders Interuniversity Institute for Biotechnology, Ghent University, K.L. Ledeganckstraat 35, 9000 Ghent, Belgium

M. KARIMI
Department of Plant Systems Biology, Flanders Interuniversity Institute for Biotechnology, Ghent University, K.L. Ledeganckstraat 35, 9000 Ghent, Belgium

C. KRYSIAK
Institute of Biochemistry and Biophysics, Polish Academy of Sciences, 5A Pawinskiego Street, 02-106 Warsaw, Poland

J.A. MCCOMB
Biological Sciences, Murdoch University, Murdoch, W.A. 6150, Australia

J.F. PETOLINO
Dow AgroSciences, 9330 Zionsville Road, Indianapolis, Indiana 46268, USA

C. QIHUA CAI
Dow AgroSciences, 9330 Zionsville Road, Indianapolis, Indiana 46268, USA

D. SIMMONDS
Eastern Cereal and Oilseed Research Centre, Agriculture and Agri-Food Canada, C.E.F. Building 21, Ottawa, Ontario K1A 0C6, Canada

D. TJOKROKUSUMO
Biological Sciences, Murdoch University, Murdoch, W.A. 6150, Australia

D. TOMES
7300 NW 62nd Avenue, P.O. 1004, Johnston, Iowa 50131, USA

W. Van de Velde
Department of Plant Systems Biology, Flanders Interuniversity Institute for Biotechnology, Ghent University, K.L. Ledeganckstraat 35, 9000 Ghent, Belgium

M. Van Montagu
Department of Plant Systems Biology, Flanders Interuniversity Institute for Biotechnology, Ghent University, K.L. Ledeganckstraat 35, 9000 Ghent, Belgium

S. von Arnold
Department of Plant Biology and Forest Genetics, Swedish University of Agricultural Sciences, Box 7080, 750 07 Uppsala, Sweden

M. Welter
Dow AgroSciences, 9330 Zionsville Road, Indianapolis, Indiana 46268, USA

S.J. Wylie
Biological Sciences, Murdoch University, Murdoch, W.A. 6150, Australia

Z. Zhao
7300 NW 62nd Avenue, P.O. 1004, Johnston, Iowa 50131, USA

1 Exclusive Rights in Life: Biotechnology, Genetic Manipulation, and Intellectual Property Rights

E.R. Gold

1.1 Introduction

Attending any biotechnology conference will confirm it. Amid all the discoveries and developments in the applied sciences we loosely group under the heading of biotechnology, it is impossible to ignore the palpable and ubiquitous presence of commerce. While investors, managers, and financial markets may not share the same enthusiasm about the inner workings of living organisms as a bench scientist, they certainly share the excitement of uncovering novel ways to make money. The key is to turn inventions and developments into a commercial product.

It is at the junction of science and commerce that the system of laws we call property and intellectual property step in. Essentially, property and intellectual property rights – including patents, trade secrets, and copyright – bridge the gap between new scientific development and commercial exploitation by packaging these developments in a way that industry can use. Property and intellectual property rights create exclusive rights to control biological matter and biotechnological innovation. Industry uses these rights to attract financing, build alliances, and, not least of all, entice scientists to participate in the project of turning a bit of knowledge into something to sell.

The application of property and intellectual property rights to biotechnological innovation is anything but simple or straightforward. Arguments about these rights abound, touching on everything from the criteria used to determine when to award them (Barton 2000) to the social and ethical implications of these rights (Knoppers et al. 1999; Gold 2000). Put 20 patent experts in a room and disagreement is sure to arise on some, often seemingly esoteric, point of law. Put those same experts together with health professionals, farmers, or nongovernmental organizations and conflict is sure to erupt.

Despite the lack of complete agreement on all aspects of property and intellectual property rights in biological matter, most of the principal strands of how these rights apply to this matter are beyond controversy. Biological matter can be controlled and the rules about when we give out those rights of control are fairly straightforward. With time, patent offices, courts and, occasionally, legislatures fill in the remaining blanks.

This chapter provides an overview of the application of property and intellectual property law to biological materials. In Section 1.2 of this chapter, we

Molecular Methods of Plant Analysis, Vol. 23
Genetic Transformation of Plants

divide innovation in the biotechnology field into two types: material and immaterial. We next, in Section 1.3, sketch out the principal property and intellectual property regimes that apply to these innovations, concentrating on those of particular relevance to industry. In the last section, Section 1.4, we review some of the challenges that property and intellectual property law faces in dealing with biological materials.

1.2 Biotechnological Innovation

From a legal point of view, biotechnological innovations have both a material and an immaterial aspect. Each molecule, cell, and organism has a physical existence. These materials can be used to produce things (e.g., silk, hormones, and spider webs), to transplant into humans (e.g., organs and stem cells), to accomplish various purposes (e.g., animals used in research), and to grow and consume (e.g., seeds, plants, and animals). In addition to this physical aspect, biotechnological innovations also have an immaterial aspect. Biological material contains, after all, information about how to make more of it or of something else. For example, a living cell or organism contains instructions and a mechanism to divide in order to produce daughter cells. DNA contains the instructions for making copies of proteins. Each of these uses of biological material is as important or more important that the actual use of the physical material itself.

The distinction between physical material and immaterial information is an important one for the law. Legal systems use different legal regimes to allocate control over and access to physical objects than they do over access to information. It is, therefore, always important to differentiate between the physical and immaterial aspects of biotechnological innovation. For example, the Sox-9 gene has both a physical existence – you can collect it in a test tube – and carries information – how to make the protein that can be used to promote bone and cartilage growth. We protect the physical embodiment of the gene through property law while we protect the information content of the gene through patents or trade secrets. In fact, US patent 6,143,878 has been issued on the use, reproduction, and sale of isolated DNA molecules containing the gene. The physical DNA molecules containing the gene are not, however, subject to the patent.

In addition to these two aspects of biotechnological innovation, there are processes that use biological material. For example, the process of making beer or wine from yeast is a process using these materials. These processes may also be subject to a property right. Returning to the Sox-9 gene, we could in theory create a process to insert that gene or cells containing that gene into someone suffering from cartilage or bone damage and seek rights to the process, either in patent or through trade secrecy protection.

1.2.1 Physical Innovations

Starting with the physical aspect of biotechnological innovation, we will survey the various kinds of biological matter that are of interest to modern biotechnology. We start with molecules, ranging from proteins to genetic sequences. We then move up to whole genomes, to cell lines, to tissues and organs and, finally, whole organisms.

1.2.1.1 DNA and Protein Molecules

DNA sequences (or corresponding RNA molecules) exist in their natural state within the cells of organisms. For most organisms, DNA contains the genetic information in that organism (the rest rely on RNA).

Apart from DNA sequences in their natural states, we can artificially create DNA molecules. For example, we can cut out a portion of DNA from the chromosome in which it usually resides corresponding to an entire gene. Alternatively, we can create numerous copies of smaller pieces of DNA, such as Expressed Sequence Tags (ESTs). If we are only interested in the DNA necessary to code for a particular protein, we can copy the edited RNA back into DNA to form complementary DNA (cDNA). This cDNA contains only the particular gene about which we are concerned.

Property rights of various sorts can exist in any of these molecules. Thus, one could own a particular piece of DNA, RNA, cDNA, or EST. We must remember, however, that property rights exist not only in the physical molecule, but in the information contained in that molecule. Usually, the value of the information contained in the DNA molecule is greater than the value of a physical copy of the molecule itself. More on this later.

The second set of molecules that are of particular interest is the set of proteins produced by an organism. Proteins are the molecules for which genes code, take on various forms and do the work of the organism. We can again separate physical proteins from the information that they contain. Unlike genetic material, the primary use of which is to carry information, proteins' primary use is to do and make things. We can thus use proteins to carry out work for us outside their normal environment. Therefore, the value of property rights in physical proteins molecules are at least as high as the value of the information they contain.

1.2.1.2 Cells

Almost every cell contains a full set of genetic information and a subset of the proteins and other ingredients to carry out life. While cells cannot usually live outside the organism in which they belong, we have developed techniques to overcome this limitation and create cell lines. We have also developed methods to keep cells alive in tissues and organs for a period of time outside of the body. We may want to do this, for example, in order to transplant an organ.

Property rights can exist in physical cells, cell lines, including stem cells, and possibly tissues and organs. As stem cells promise everything from replacement brain or heart cells to new organs, the value of particular, tangible, stem cells may be high.

1.2.1.3 Whole Organisms

Property rights in complex higher life forms – plants and animals other than unicellular organisms – are common. After all, we are used to owning particular cows, pigs, and other farm and domestic animals as well as a variety of plants. Nevertheless, ownership of animals in particular gives rise to ethical concern over the treatment of animals both in agriculture and in research. With the advent of transgenic animals, concern has already been raised over the suffering of animals modified so as to suffer from some disability (European Patent Office Board of Appeal 1990).

1.2.2 Information and Other Intangibles

As discussed above, it is not only the physical aspect of biotechnological innovations that are interesting from a scientific and commercial point of view, but the immaterial aspects of these inventions. Biological matter such as DNA sequences contain information that is valuable both for developing new therapies for disease and for creating organisms with novel characteristics. Biological matter also contains information about the organism to which it is derived. For example, a blood sample can help identify an individual as having committed a crime or of being at a higher risk for developing a particular illness. In this section, we review the immaterial nature of biotechnological innovations.

1.2.2.1 DNA Sequences and Cells

Because DNA molecules contain genetic information, they are particularly important and valuable. As stated earlier, we are likely to be more interested in the information contained in a particular cDNA – the information about how to construct a particular protein – than in a single physical cDNA molecule. This means that the property rights in the information contained in these molecules – that is, the right to copy (reproduce), the right to use, and the right to sell access to this information – is often of greater value than rights in the molecules themselves.

Similarly, the right to reproduce cell lines can often be as valuable as holding the cell lines in your possession. Again, the information contained in the cells – for example, the instructions they contain on how to reproduce themselves – are subject to property rights.

1.2.2.2 Processes Using Biological Matter

In addition to biological matter, we can grant property rights to processes that make use of this matter. We already put biological matter to work making beer, wine, cheese and pharmaceuticals. While some of these processes are very old, some depend on modern biotechnology and the manipulation of genetic material within organisms to create them.

In addition to manufacturing things, we can use biological matter to perform certain functions. For example, we can use bacteria to bioremediate polluted soil. The process of using these bacteria, whether genetically modified or not, constitutes valuable information. Similarly, ways to conduct gene therapy using vectors and DNA sequences is information of value.

All of these processes can, at least in theory, be subject to property rights. While these property rights do not attach to the physical molecules or cells themselves, they relate to the use of these molecules and cells.

1.2.2.3 Bioinformatics

Both genomes and proteomes – being sets of information – are by their nature intangible. Yet, despite not having a dual physical material and immaterial aspect, genomes and proteomes can be looked upon as two different types of sets of information. The first is raw genomic or proteomic information: the sequence of the DNA as it exists in a representative organism of a species or a list of proteins that exist within that organism. The second type of information is the organization of this material into a database. This involves the creation of a database that supports the retrieval and manipulation of useful information about parts of the genome or proteome (such as identification of individual genes, the comparison of two sequences, models of the structure of the protein produced from a gene, etc.).

Both the raw genome and proteome on the one hand, and the organized genome and proteome on the other are valuable for different reasons. This value can be protected through different property rights. We can, for example, grant property rights over an entire genome or over the organization of that genome within a database. In the first case, we are interested in capturing the value implicit in the information that is the genome itself; in the second, in the value implicit in being able to compare and query that information.

1.2.3 Summary

Biology comes to property law in two forms and many levels of complexity. Its first form, as physical matter, includes particular molecules of DNA, particular cells, and particular plants. The second form is the information contained in the biological matter such as the code contained in a DNA molecule or the use of bacterium in bioremediation.

The value of each different type of biological entity varies according to the ways in which it is used. For some material, such as DNA, the value lies primarily in its information content. For others, such as proteins, the value may lie either in the physical molecule or in its use.

As we will see below, different types of property regimes apply to these different forms of biological matter.

1.3 Introduction to Intellectual Property Rights

When someone creates something new, whether in biology, electronics, or in mousetrap building, he or she often tries to capture the commercial value of that innovation by seeking a property right. Property rights provide the innovator with control over how the innovation is used and by whom. There are two general types of property rights that apply to these innovations: personal property rights and intellectual property rights. As we have seen above, innovations have both physical and informational components. In general, personal property rights protect the former while intellectual property rights give control over information.

Before examining the differences between personal and intellectual property rights, it is important to understand what property rights actually are. Too often common understandings of "property" or "ownership" are wrong, leading to crucial misunderstandings. We discuss this below.

1.3.1 Exclusive Rights vs. Rights to Things

Property rights are, in whatever form they take, essentially negative rights to prevent other people from doing something. That is, all property rights, whether in land or to food, patents, copyrights, or plant variety protection, give to the holder of that right the ability to prevent others from carrying on certain activities such as using, transferring, or copying the subject matter of the property right. Thus, all property rights are "negative" rights in that they can only be used to prevent others from engaging in a certain set of activity.

The reason that property rights are negative rights is simple enough to understand. First, we do not have rights in things, but rather against other people. For example, I have no right to use the subject of one of my property rights; rather, I only have the right to prevent others from using it. Second, property rights are always subject to someone else having better rights. A building owner does not, for example, have the right to use that building in any way he or she chooses. This is because his or her use is subject, for example, to zoning by-laws as well as to the rights of neighbors. Thus, having a property right is no guarantee that one can actually use or sell the property in question. This is true no matter what kind of property one has.

All this means that, from a purely legal point of view, all property rights consist of the ability to prevent others from undertaking certain action. In this way, property rights are nothing more than vetoes over the activities of others. Like a veto, a property right gives no affirmative ability to do something; it merely allows us to prevent activity that we do not like or want.

1.3.2 Property and Intellectual Property Rights

Personal property rights arise from general legal principles surrounding rights to use particular things. Most often, personal property rights relate to physical objects such as particular flasks, cars, or hats. These are the familiar forms of personal property. However, personal property can also relate to biological materials, such as particular cells, particular molecules of DNA, or even particular laboratory animals. What is important to remember is that, by their nature, personal property rights almost always relate only to particular, identifiable things.

Intellectual property rights (IPRs) exist because governments around the world have passed special legislation to protect certain forms of immaterial ideas or ways of expressing ideas. These rights apply to certain forms of immaterial "objects". As each intellectual property system (the most important being, for biotechnological innovations, trade secrets, patents, copyright and plant breeders rights) differs significantly from the others, I describe each in turn.

1.3.3 Trade Secrets

Trade secret protection involves maintaining control over the disclosure and use of information. The person claiming trade secrecy protection provides access to information only to those who agree to the keep that information secret. Trade secrecy protection (also called the protection of confidential information) permits those with business data or technical results (including negative results) to prevent others from using that information. If someone wrongly discloses information that is a trade secret, the person to whom that information belongs can ask a court to award damages against the disclosing party and to prevent the receiving party from using the information.

Biological matter provides a good breeding ground of trade secrets. This is because biological matter is often difficult to reproduce without access to the original sample or original information on where to collect another sample. This makes it an excellent candidate for trade secrecy protection since there is little practical chance that another researcher or company will independently create it. Trade secrecy protection has one very attractive characteristic: it is potentially indefinite in duration. As long as the information is actually kept secret, it continues to be a trade secret.

1.3.3.1 Subject Matter

Trade secrecy protection is only available in respect of identifiable information that is actually maintained as a secret. This means that the information can never be disclosed except where it is understood that it is confidential and cannot be further disclosed by the receiving party. Different jurisdictions take slightly different approaches to what information is considered confidential. For example, under the Uniform Trade Secrets Act and the Restatement of Unfair Competition in the United States, the information must be of value, not generally known or readily ascertainable by proper means, and be the subject of efforts reasonable under the circumstances to maintain its secrecy. Under the European Patent Convention, the trade secret must be fixed in a tangible medium.

1.3.3.2 Requirements

Trade secret protection only applies to information so long as it is actually kept secret. If the information is publicly disclosed or independently created, the protection ends. That is, as soon as the person claiming trade secrecy protection loses control over the information (unless by improper means), the information is no longer confidential and ceases to be protected.

Provided that the holder of the secret information maintains its confidentiality, the protection offered by trade secrecy is theoretically unlimited in terms of time. This contrasts with other forms of intellectual property rights, such as patents and copyrights that are of limited duration. This aspect of trade secrecy protection makes it very attractive. To the extent that it is possible to maintain the secrecy of a discovery – such as a rare bacterium or complex market information – companies will often rely heavily on this form of protection.

1.3.4 Patents

A patent provides its holder with the ability to prevent all others from making, using, selling or importing an invention. We grant patents in inventions that are new, nonobvious, and that have an industrial application (Caulfield et al. 2000). Something is new if it has not been described before in public (subject to a grace period in certain countries). An invention is nonobvious if it would not have been entirely obvious to a researcher in the field to have created the invention taking into account the current state of publicly available knowledge in that field. An invention has an industrial application if it has some practical application.

Patents are governed by a web of international intellectual property agreements from the *Paris Convention for the Protection of Industrial Property* (which sets out the basic rules for priority of patents and national treatment), the *Patent Cooperation Treaty* (setting out an international framework for the searching and examination of patents), and the *Patent Law Treaty* (which is not yet in force, but supplements the procedure set out in the *Patent Cooperation Treaty* for the filing of patent applications). International trade agreements, most notably under the World Trade Organization's *Agreement on Trade Related Aspects of Intellectual Property Rights* (TRIPs) also establish rules with respect to patent protection (Gold 2001a). In addition to these broad international agreements, several regions have established regional patent conventions such as in Europe, in the Newly Independent States, Eurasia and in Africa. These regional conventions establish procedures to obtain patents in the member countries through a unified regional procedure.

Despite these international agreements, patents are, at bottom, national in character. This means that a US patent is good only in the United States; it has no effect outside that country. Therefore, if an inventor only has a US patent, he or she cannot prevent someone else from making, using, selling, or importing that invention into, for example, Germany or Thailand. The situation is the same for all other countries.

1.3.4.1 Subject Matter

Since at least the 1980s, inventors have sought to protect the practical knowledge they gained about biological matter through patent law. Twenty years is a short period for any body of law to develop, especially one dealing with such a complex and ever-changing technology as is modern biotechnology. Nevertheless, borrowing from principles developed over the past century or more, patent law has adapted itself – not without criticism – to the modern world of biotechnology. Of course, more development is needed, but the basic outlines of patent law's application to biotechnological innovation is well known.

Patents are granted in inventions. In general terms, an invention is a practical application of knowledge to be used in an industrial or commercial setting. In contrast to the fine arts, inventions must be useful rather than aesthetic, thought provoking, or pleasurable. It is, of course, impossible to give a definitive definition of invention for the simple reason that we cannot imagine what will be invented in the future. Nevertheless, patent law uses several tests to determine whether something is an invention. I discuss these below.

1.3.4.1.1 Invention vs. Discovery

One of the critical moments in research, from a patent law point of view, is when the researcher moves from the creation of new knowledge to the creation of an invention. At the heart of this transition lie the concepts of invention and of discovery. Inventions can be patented while discoveries cannot.

In order to help understand this distinction, consider a gene from a cow. The gene as it exists in the cow's body is not patentable for the simple reason that the gene has existed there for a long time. However, one gene mixed in with the thousands of others in a particular cow's body is of little interest to anyone save the cow. To be useful for research or commercial purposes, the gene must be isolated and purified. That is, it must be removed from the cow's body, separated from the rest of the cow's DNA, and reproduced many, many times.

Unlike genes within the cow, isolated genes are patentable if they otherwise meet the criteria of being new, have an inventive step and have an industrial application (Gold 2001b). This is because, in the course of history, genes have never come neatly packaged in isolated and purified form. It takes human intervention to put the gene in this form.

In theory, any biological matter isolated from a human, animal, or plant can be patented. Thus, according to general patent theory, ESTs, proteins, molecules, and cells isolated from living organisms can be patented provided that they otherwise are new, contain an inventive step and have an industrial application. The mere fact that the "invention" is a distilled version of what nature produces does not disqualify the invention from being patented.

1.3.4.1.2 Exclusions

Despite the theory, not all biological matter and their uses are, in fact, patentable in all countries. Many countries do not, for example, grant patents on plants and animals and some do not grant patents over genes. Several developing countries claim that these patents encourage companies from the developed world to patent their biological inheritance (Ekpere 2000). This, they claim, not only undermines the economic development of developing countries, but robs their populations and communities of traditional knowledge.

Certain processes using biological materials are similarly unpatentable. For example, in Europe and in Canada, medical treatments (and, in Europe, diagnostic procedures) performed on a live body cannot be patented even though some of the biological material used in these procedures may be patented (Gold 2001a).

The exclusions discussed so far have dealt with a certain category of inventions. Some countries also exclude inventions from patent coverage based on their particular nature. Once such limit is the "ordre public" or morality exception to patent law. This exclusion allows patent offices to reject patent applications that otherwise meet the novelty, nonobviousness, and industrial application requirements because the invention itself or, more precisely, the commercial use of that invention, would violate "ordre public" or morality. An invention that violates ordre public would be one where the commercial use of the patented invention causes significant public unrest and political disorder. Morality refers to generally accepted moral norms within the particular society. For example, patents that cover embryos that are likely to grow to term are often thought to violate morality (Gold and Gallochat 2001). The European Patent Office and the national patent offices within Europe and Asia are able to reject patents on the basis of a breach of ordre public or morality while those in the United States, Australia and Canada cannot.

1.3.4.2 Requirements

While trade secrecy protection (as well as copyright) arise without any explicit official act by the secret's originator, an inventor wishing to gain patent protection must actually file a patent application in every country in which he or she wishes to obtain a patent. This may sound onerous – it often is – but there are numerous international agreements to smooth out the difficulties that would otherwise be encountered in applying for patents in many different countries with different versions of patent law. In addition, countries are negotiating further international agreements to facilitate the application, grant, and enforcement of patent rights around the world.

To obtain a patent, an inventor (or his or her employer) must apply for it from the desired national or regional patent offices. These offices must be satisfied that the invention meets that country's requirements for obtaining a patent in that country. Thankfully, the basic requirements for patentability are set out in several of the international agreements described earlier (Gold 2001a). While countries interpret these requirements slightly differently from one another, there is much consistency between them.

According to the international agreements, patent offices are to apply three substantive and one procedural criteria to determine whether to award a patent. The three substantive criteria are novelty, inventive step, and industrial application. The procedural criterion is the requirement of describing the invention.

1.3.4.2.1 Substantive Criteria

Before a patent office grants an applicant a patent, it must ensure that the invention meets certain substantive requirements. These requirements weed out innovations that are not practical or do not bring new information to the public. In other words, they act as a screening device to ensure that not too many rights are granted.

1.3.4.2.1.1 Novelty Since patents are granted over new creations, the first criterion that patent offices apply in assessing a patent application is that the invention is new, or novel. This sounds straightforward, but it is not always easy to determine whether a particular invention is, in fact, new. Patent examiners search through their records and publicly available records to determine whether the same invention has been disclosed elsewhere in a single source. If so, the invention is not new. If not, the invention satisfies this first criterion.

1.3.4.2.1.2 Inventive Step (Nonobviousness) Even if an invention is new in the sense that it has never been described before in the literature, it may not be worth protecting. This is because it may be so obvious that anybody with skill in the relevant discipline may have been able to create it. As there is no sense in rewarding a person for developing something than anyone could have created, patent examiners look at what is known in the relevant disciplines to determine whether the inventor had exercised some creativity in developing the invention. While the inventor need not display a large amount of creativity, he or she must bring some creative element to the invention.

In many countries, this is a relatively easy step to overcome. In others, such as in Europe, it is more difficult. This is because these patent offices require some evidence of function in determining whether an invention has an inventive step.

1.3.4.2.1.3 Industrial Application (Utility) The most difficult of the three criteria to meet is usually that of industrial application. To understand the concept, consider the following example. A CD player is useful to play music, but is also useful as a paperweight or as a doorstop. Given the wide variety of uses to which an invention could be put, we can apply the industrial application standard either restrictively or liberally. The United States Patent and Trademark Office (USPTO) – where the industrial application standard is called the utility standard – takes a more restrictive approach to industrial application than does, for example, the European Patent Office.

In 2001, the USPTO issued guidelines on how it will apply the utility standard to biotechnological inventions. These guidelines state that an invention can meet the utility requirement by possessing a specific, substantial, and credible utility (United States of America 2001). This can be demonstrated in one of two ways.

The first way in which an invention can be shown to be useful is where someone with knowledge of the area would immediately recognize that the invention is useful. That is, where the utility of the invention is so clear that anyone in the field would see it, the invention is deemed to be useful. Many gene-based inventions will not meet this test since people in the field will not know, in advance, the specific ways in which the invention can be used. In this case, the invention would need to qualify as useful in the second way.

The second way for an invention to be demonstrated as useful is if the inventor shows that the invention has each of a specific, a substantial, and a credible utility. To have a specific utility, the invention must be useful in a way that is unique to it. So, for example, while many CD players can play music, the claimed CD player has the special ability to better resist outside vibration. For an invention to have a substantial utility, it must have a real world, commercial utility. Neither basic research nor general statements that the invention may be useful in curing disease count as a substantial utility, the first because it is not a commercial utility and the second because there is no substantive reason given why the invention should be any better than anything else. Similarly, the substantial utility of the CD player cannot be that it can be used as a doorstop since many objects can equally be used as doorstops. A credible utility is one that would be believable to someone in the field of research based on evidence supplied. This means that there must be some basis in published material or in general knowledge in the field that would lead someone to conclude that the claimed utility could actually be true.

The general industrial application standard used under the European Patent Convention is that an invention has an industrial application if it can either be manufactured or it has some plausible industrial use. This standard is fairly low. Any activity that belongs to the useful or practical arts, as opposed to the aesthetic arts, is sufficient to meet the industrial application requirement in Europe. In 1998, the European Community adopted the *Directive on the legal protection of biotechnological inventions* (the *Directive*). The *Directive* states that genes, DNA sequences, and proteins are clearly patentable provided that these substances meet the general requirements of novelty, inventive step and industrial application. The *Directive* also states that, in respect of human genetic sequences, an inventor must indicate the function of the sequence (in most cases, the protein for which it codes and possibly the function of that protein) in order for it to be considered to have an industrial application. Unfortunately, the enforceable portion of the *Directive* is ambiguous on the application of the industrial application standard to genes of nonhuman origin (presumably, even if identical to a human gene).

1.3.4.2.2 Procedural Criterion: Disclosure

An inventor who creates an invention that is novel, involves an inventive step and has an industrial application is entitled, subject to an objection on the grounds of "ordre public" or morality, to a patent provided that he or she complies with the procedural requirements for filing a patent. The primary of these requirements is the obligation to fully describe the invention in the patent application.

The obligation to disclose the invention is critical to patent law. It is what differentiates it from trade secret protection. As the application becomes, in most cases, public 18 months following its initial filing, the inventor will lose the ability to protect the invention as a trade secret. This is intended by patent law: in return for the greater rights that the inventor obtains through a patent, he or she gives up any claim to secrecy over the invention. The public thus gains access to the invention so as to learn from it even if the public cannot practice the invention without violating the patent right.

An invention must be disclosed in sufficient detail so as to allow someone with skill in the relevant disciplines to make or use it. Thus, the inventor cannot hide so many details about the invention that it becomes difficult to recreate.

1.3.4.3 Remedies

There are two principal remedies available to an inventor against someone who makes, uses, sells or imports an invention without permission. The first is damages for having violated the patent. Different countries calculate the level of damages differently, but generally the damages are based either on the loss suffered by the patent holder or the gain made by the person wrongfully using the invention. The second is an injunction, or order preventing the future wrongful use of the invention.

Not every use of an invention is a breach of a patent. When the patent holder sells an object encompassing an invention, he or she cannot prevent use of that object even if it is resold. That is, the patent holder exhausts his or her rights on the first sale of an object incorporating the invention. Similarly, the patent holder may give permission to someone to use the invention. This is called a license. As long as the user complies with the terms of the permission granted, there is no violation of the patent.

1.3.5 Copyright and Database Protection

While trade secrets and patents protect the information contained in biological matter directly, copyright protects the manner in which this information is stored, organized, retrieved, and modeled. With the great rise in genetic information, databases, search tools, and modeling tools that enable researchers to identify promising genes and the proteins they produce have become very valuable. Copyright provides the creators of these goods with the right to prevent all others from copying them, but does not cover the ideas conveyed in these works. That is, copyright provides no protection over the information contained in a database (the DNA sequences or proteins). It does, however, protect the way the database is structured and how to search and use the information contained in it.

Copyright protection has become a valuable property right to those who have accumulated large amounts of information about biological material. For example, Celera Genomics, the private company that had sequenced the human genome in competition with the public Human Genome Project, is charging a fee for access to its database of raw sequence data for use in commercial research. Given that the organization of information is often as valuable as the underlying information itself, copyright will likely be an important way that those working in the biotechnology field protect their work.

In some countries, such as in the Member States of the European Community, protection over databases is provided through legislation separate from copyright. Nevertheless, the general principles governing this form of protection are similar to that of copyright.

Copyright in databases lasts for anywhere between 50 and 75 years after the last one to die of the authors of the database.

1.3.5.1 Subject Matter

Copyright protects the form rather than the substance of an expression. Copyright gives its holder the right to prevent others from copying the way in which the holder has expressed an idea. Thus, the author of a novel has a copyright in the way he or she expressed the story in terms of character, setting, and sequence of words used. The author receives no right, however, in the idea of a novel, characters or setting or even of the general storyline (star-crossed lovers, honor and sacrifice, etc.) behind the novel.

Similarly, because raw genetic information is substance and not expression, copyright does not protect it. In contrast, the selection of this raw information and the selection of how to store it and retrieve it involve personal choices of how to express that information. They are thus subject to copyright protection.

1.3.5.2 Requirements

Copyright applies to all expressions of a work. While the origins of copyright law may lie in the publication of literature, it has since been made to apply to all forms of art and music and even commercial expressions such as commercial jingles, computer software, and office manuals. In short, anything that consists of an original expression, as long as it is in a fixed form (so that it can be identified), is subject to copyright protection. Remember, however, that the content of that expression – the ideas contained in the work – are free to use. The difference between expression and content is fairly easy to see in the following example. A cook may write a cookbook containing a new and particularly delicious recipe. Copyright prevents others from copying that recipe, but not from making the novel dish. Thus, the form of expression – the recipe – is protected while the idea behind it – the dish – is not.

Because of its preoccupation with creativity, copyright does not extend to the expression of ideas where there are only a limited number of ways of expressing that idea. Where the idea itself calls for a particular expression (consider $E = MC^2$), the expression, as opposed to the idea behind the expression, is not considered sufficiently creative to merit copyright protection.

Given that organizing genetic information involves choice and skill, the construction of a database of genetic information is subject to copyright (or database protection where this protection exists). This means that, while the information contained in the database is free to be used, the actual use of the database and its accompanying tools are subject to copyright.

On the other hand, database protection gives the holder limited protection over the raw information extracted from the database even though that information does not represent any expression. Thus, the owner of a database right could prevent others from copying extracted information for certain purposes.

There are no formal requirements to gaining copyright protection. By international agreement, the moment the author expresses something, that expression is protected by copyright. Nevertheless, there are some procedural advantages to registering a copyrighted work with national copyright offices.

1.3.5.3 Remedies

As with patents, a copyright holder can seek damages or an injunction against anyone who copies the expression.

1.3.6 Plant Variety Protection

Many countries have enacted special legislation to protect the interests of those who breed new plant varieties. (Plant varieties are plants below the species level that exhibit a particular defined characteristic.) At the international level, this protection is provided by the International Convention for the Protection of New Varieties of Plants. The Convention and national implementing legislation provide plants breeders with the ability to exclude others from reproducing, selling, importing, or exporting plant varieties and, in some countries, plant varieties derived from protected plant varieties.

Plant variety protection is often accompanied by a so-called farmers' privilege that permits farmers to store and use seeds collected from a plant of the protected variety.

Plant variety protection lasts for at least 15 years and, depending on the country, often significantly more.

1.3.6.1 Subject Matter

Plant variety protection provides breeders with protection over the botanical taxon of the lowest known rank. Varieties must display at least one stable and defining characteristic to warrant protection. Flowers and nonhybrid plants can often be protected in this way.

In general, plant variety protection is not suitable for biotechnological innovation. This is because the technologies involved are seldom if ever limited to a single plant variety. For example, it would be easy for an individual to circumvent the protection by simply introducing the gene with the characteristic in question into a different variety. As this new variety would not fall within the protection offered, the protection is of little use to its holder.

1.3.6.2 Requirements

A plant variety must generally be new, distinct, uniform and stable in order to be protected. A variety is new if it has not been previously sold for exploitation. This requirement is subject to a 1-year grace period in the country in which protection is sought and a 4–6 year grace period in other countries. To be distinct, the variety must be clearly distinguishable from other known varieties. A variety is uniform if, given natural differences, the varieties remains sufficiently uniform in the identifying characteristic. A variety is stable if it maintains the identifying characteristic after repeated propagation.

1.4 Challenges

There are obvious challenges to simply applying the various property regimes described above to the different aspects of biological matter. In addition to these technical difficulties, lie others that are often more difficult to address. These other difficulties range from the abstract concerns over the morality of patenting life, to very practical concerns regarding access to new health care procedures and to plant platform technologies. It is to these difficulties that this section is dedicated.

The morality of having property rights in living organisms is not a topic that is easy to resolve. Different philosophic and religious traditions lead to different positions about the appropriateness of permitting an individual to control how the remaining members of society use a particular living organism or part of an organism. Many intellectual property practitioners believe this debate misses the point. They point out that intellectual property rights are not about the ownership of specific organisms, but more abstract rights to reproduce them and use them in a commercial setting. On the other hand, those who have moral objections to granting property rights in respect of biological matter reject this argument as missing the central point that no human being ought to have control over the essence of a living organism including control over who has access to it for which reasons. As will be obvious to the reader, these debates are endless and not subject to easy resolution.

As opposed to these more abstract concerns relating to the application of intellectual property rights to biological matter, lie more utilitarian concerns regarding the social and economic effects of granting these rights. There are many such concerns and space only allows for a selection of these to be presented. In what follows, we examine two issues.

1.4.1 Incentive vs. Access

One of the challenges of intellectual property law is to find the appropriate balance between the rights granted to the creators of new things against the rights of the rest of us who would like to use or make those new things. In other words, intellectual property regimes seek to determine the amount and kind of incentive to provide inventors while maintaining public access to innovation. This is by no means easy and it is by no means obvious that there exists a single balance appropriate to all inventions and all knowledge. As pointed out by one senior US judge: "Indeed, commentators complain that a single patent policy and patent law are unsuited to the range of scientific and commercial activity in today's economy" (Festo Corporation vs. Shoketsu Kinzoku Kogyo Kabushiki Co., Ltd. 2000).

1.4.1.1 Justification for Property Rights

There is, in reality, no one single justification for all property rights. Theories abound, from wealth creation (Posner 1981), to personal development (Hegel 1852; Radin 1982), to dessert (Locke 1980) to explain why we grant property rights. Attempting to tackle the big issue of why we grant property rights and why we grant particular kinds of property rights has filled volumes of text with no consensus reached. Instead of entering into this general debate, we will bring it down a level and examine the justification for intellectual property regimes as they apply to practical innovations. We thus leave out ordinary property rights as well as copyright and trademarks from this discussion.

Given the large range of justifications given for the award of property rights in general, it may come as a surprise that there is a consensus on why we award property rights in the applied arts. Since the earliest days of the modern patent system in the late 18th century, everyone has agreed that the purpose of this system is to achieve the overall social good by encouraging inventors to disclose their knowledge and to bring that knowledge to the public.

To understand this, we need to step back and look at what inventors would do in the absence of a patent: many would try to protect their knowledge as a trade secret. The problem is that trade secrecy protection can potentially lead to a hoarding of information in the hands of a few researchers or companies. Science is best accomplished, it is thought, through the free exchange of information. Trade secrecy undermines this free exchange by encouraging researchers and their sponsoring companies to take steps to maintain the confidentiality of their inventions. This may lead to stagnation since all researchers and companies will find their research paths blocked through lack of access to others' inventions. For example, in the seed industry, different companies may improve seed stock. If each keeps its advances secret, no one will practicably be able to put all of these improvements together in the same seed.

It is to counter this negative effect of trade secrecy that patents were born. Essentially, a patent provides its holder with a limited term, stable monopoly (that is, a more valuable set of rights than is provided by trade secrets) in return for disclosure of the invention. The patent system is designed to induce inventors to abandon their claim to trade secrecy protection so that the public gains the benefits of the invention. Patents have a second, equally important, advantage. Because patents provide certainty, they permit inventors to raise money to invest in the further development of their inventions and to bring those inventions to the market. Thus, in a perfectly working patent system, we should see both an increase in research due to disclosure and an increase in products brought forward to the market due to the stability and value of the patent. Accordingly, patents ought to provide inventors with just enough – but not more – of an incentive to disclose and bring their products to market.

Of course, this simplifies the justification of patent law a little too much. Patents are, as stated earlier, designed to achieve the overall social good. Thus, simply encouraging research and the bringing of products to market does not necessarily achieve this goal. For example, encouraging the private development of a bomb or biological weapon is unlikely to serve the public good. Thus, most countries include an "ordre public" or morality clause as discussed earlier to exclude inventions that undermine morality from patent protection. Similarly, if the patent system encourages the bringing of, say, a life-saving invention to the market, but at such a high price that it cannot be accessed by those needing it, society may be the loser. This is because, with a patent, no one other than the inventor is entitled to make or use the invention for 20 years; if the invention had instead been protected by trade secret, someone else may have been able to recreate it within that time, leading to competition.

1.4.1.2 Economic Reality

One may have thought that, given our 200 years of experience with the patent system, we would know whether it actually serves the goal of attaining the overall social good. Unfortunately, our knowledge about the actual economic effects of patents is modest, particularly in the biotechnology field (Gold et al. 2002). Economic studies have alternatively found positive and negative effects of patenting. Some studies suggest that industry prefers trade secrecy protection over that given by patents (Cohen et al. 2000). In addition, despite the theory, industry often seeks patent protection not to provide them with a stable right, but to trade for access to other people's patents.

None of this is to say that patents do not necessarily work as designed. The best we can conclude is that we simply do not know. This becomes more of a problem for biotechnology than for other disciplines since the price of getting the patent system wrong is that we overemphasize the incentive part of the balance at the cost of unduly limiting access to innovation by both researchers and end-users in the health and agricultural sectors.

1.4.2 Fairness to Providers of Biological Matter

A second issue that has increasingly gained attention is whether those who commercialize technology based on biological matter have an obligation to share the proceeds from the invention with those who provided the biological matter (Human Genome Organization Ethics Committee 2000). This can arise both with respect to health care, where individuals will provide tissues and genetic material, and in agriculture where industry may take biological samples from particular countries or that fall within the traditional knowledge of certain aboriginal groups or other populations. The sharing can take the form of monetary payments, but can equally be accomplished through the provision of health or agricultural infrastructure or knowledge.

While the existence of an obligation to share benefits is not directly connected to intellectual property – after all, this obligation can arise even if one has not protected the invention – it is most likely to arise where the innovation has been protected. This is because almost anyone commercializing a product will have sought some form of intellectual property protection. As they currently stand, however, no intellectual property regime requires creators to share benefits with sample donors. This is because the donation of biological matter is not considered to be part of the creation of the invention that is, after all, nothing more than knowledge derived from the matter. On the other hand, but for access to the biological matter, the inventors would never have made the invention.

1.4.2.1 Rights to Biological Matter

Under the 1992 *Convention on Biological Diversity*, each country has the right to determine the conditions upon which it will provide access to its biological resources. Some countries have exercised this right by entering into agreements with industry with respect to providing access to these resources in return for royalty payments and the provision of scientific training and infrastructure.

There has been an increasing call for the sharing of benefits arising out of biomedical science with the individuals who donated tissues and genetic material to researchers. The Ethics Committee of the Human Genome Organization called, for example, on those who commercialize biomedical innovation to share between 1 and 3% of their profits with the communities from which the samples were donated (Human Genome Organization Ethics Committee 2000).

A lawsuit has been launched in the United States with respect to the failure of a research institution to share the benefits of their patent with the families who donated the tissue necessary for the research. In an earlier case in California, in which a patient alleged that his doctor had removed his spleen knowing that it contained commercially valuable molecules, the court did not find that the man had "owned" his spleen (Moore vs. The Regents of the University of California 1990). No argument was presented in that case, however, about an obligation to share benefits.

1.5 Conclusion

As the biological sciences move forward, they raise not only more scientific questions, but questions of law and ethics as well. There are certain things that are clear, however. It is now beyond dispute that property rights apply to the products of biological research. We have already determined that property rights exist in particular pieces of biological matter – DNA molecules, cells and animals – and in ways of using and making this matter. There is no doubt that, at least in the developed world, DNA sequences and cells can be patented whether they are of plant, animal, or human origin. What is less clear is the scope of those rights, limits we will impose on patents to protect fundamental ethical principles, and the effects that these rights will have on innovation, on health care systems, and on agriculture.

Once we move away from lay understandings of property rights, we see that there are, in fact, a multitude of different property rights that apply to a multitude of different aspects of biotechnological innovation. We have seen that ordinary property rights exist in the physical aspects of biological matter while intellectual property rights exist with respect to the uses and the right to reproduce that matter.

Given the complexity of both the biological sciences and the property regimes that apply to them, one is apt to feel overwhelmed. Nevertheless, once the central concept of what a property right is – a veto over the activity of all others – the application of the different versions of property law becomes clearer.

References

Barton J (2000) Intellectual property rights: reforming the patent system. Science 287:1933–1934

Caulfield TA, Gold ER, Cho M (2000) Patenting human genetic material: refocusing the debate. Nat Rev Genet 1:227–231

Cohen WM, Nelson RR, Walsh JP (2000) Protecting their intellectual assets: appropriability conditions and why US manufacturing firms patent (or not). National Bureau of Economic Research, Cambridge, MA

Ekpere JA (2000) OAU model law: African model legislation for the protection of the right of local communities, farmers and breeders, and for the regulation of access to biological resources. Organization of African Unity, Addis Ababa, Ethiopia

European Patent Office Board of Appeal (1990) T19/90

Festo Corporation v. Shoketsu Kinzoku Kogyo Kabushiki Co, Ltd (2000) 234 F.3d 558

Gold ER (2000) Moving the gene patent debate forward. Nat Biotech 18:1319

Gold ER (2001a) Patenting life forms: an international comparison. Canadian Biotechnology Advisory Committee, Ottawa, Canada

Gold ER (2001b) Gene patenting. Canadian Biotechnology Advisory Committee, Ottawa, Canada

Gold ER, Castle D, Cloutier LM, Daar AS, Smith PJ (2002) Needed: models of biotechnology intellectual property. Trends Biotech 20:327–329

Gold ER, Gallochat A (2001) The European *Biotech Directive*: past as prologue. Eur Law J 7: 328–366

Hegel GWF (1952) Philosophy of right. Knox TM (translator). Oxford University Press, Oxford

Human Genome Organization Ethics Committee (2000) Genetic benefit sharing. Science 290:49

Knoppers BM, Hirtle M, Glass KC (1999) Genetic technologies: commercialization of genetic research and public policy. Science 286:2277–2278

Locke J (1980) Second treatise of government. In: Macpherson CB (ed) Hackett Publ, Cambridge

Moore vs. The Regents of the University of California (2000) 51 Cal.3d 120

Posner RA (1981) The economics of justice. Harvard University Press, Cambridge, MA

Radin MJ (1982) Property and personhood. Stan L Rev 34:957–1015

United States of America (2001) Patent office's utility examination guidelines. Federal Register 66: 1097–1099

2 *Agrobacterium rhizogenes*-Mediated Transformation of Plants

W. Van de Velde, M. Karimi, G. den Herder, M. Van Montagu,
M. Holsters, and S. Goormachtig

2.1 Introduction

The soil bacterium *Agrobacterium rhizogenes* generates the "hairy root" syndrome in infected plants and is characterized by the neoplastic outgrowth of roots (Riker et al. 1930). The molecular basis of this phenomenon is the transfer and integration of a specific part of the root-inducing (Ri) plasmid of *A. rhizogenes*, called "transfer DNA" (T-DNA) into the genome of plant cells (Chilton et al. 1982; White et al. 1982; Willmitzer et al. 1982). The initial molecular events resemble those involved in the induction of the crown gall disease by *Agrobacterium tumefaciens*, its close relative (reviewed in Zupan et al. 2000). In the latter case, the T-DNA is located on the tumor-inducing (Ti) plasmid.

In addition to the Ti or Ri T-DNA regions, a virulence (*vir*) region is also present on the Ti or Ri plasmids. The gene products of the *vir* genes are responsible for the generation of the T-DNA intermediate transfer complex and its subsequent transfer to plant cells. This process is triggered by exudates of wounded tissues, mostly phenolic compounds, such as acetosyringone, which induce the *vir* genes. Upon stimulation, a single-stranded (ss) copy of the T-DNA is generated, which is specified by two direct repeat (25 bp) borders, the right (R) and left (L) borders. Subsequently, VirE proteins covalently bind to this ss copy, forming the intermediate transfer complex or T-complex. The VirE proteins protect the ss T-DNA strand from degradation. Besides the VirE proteins, a single VirD2 protein is coupled to the 5′ terminus of the T-DNA strand. The T-complex is subsequently transferred through the bacterial envelope to the plant cell via the type IV secretion system. This system consists of 11 proteins encoded by the *virB* operon and the *virD4* gene, also collectively referred to as the T-transporter. Because both VirD2 and VirE2 proteins contain nuclear localization signals, the T-complex is finally targeted to the nucleus. After transfer, the ss T-DNA is made double-stranded and is integrated into the plant chromosome, probably via illegitimate recombination (De Buck et al. 1999).

Expression of genes from the Ri T-DNA region results in the formation of transformed roots, which are characterized by uncontrolled growth and abundant branching. By insertional mutagenesis four root loci (*rolA*, *rolB*, *rolC*, and *rolD*) have been identified on the Ri T-DNA that are responsible for the hairy root phenotype (White et al. 1985). The Rol proteins probably cause the ontogeny of hairy roots by changing the hormone balance or sensitivity towards auxins and cytokinins (reviewed in Christey 2001). The exact mecha-

Molecular Methods of Plant Analysis, Vol. 23
Genetic Transformation of Plants

Table 2.1. *Agrobacterium rhizogenes* strains most frequently used for *Agrobacterium*-mediated transformation

A. rhizogenes strain	Ri plasmid type	Reference
A4	Agropine	Moore et al. (1979); Petit et al. (1983)
A4T (=C58C1-pRiA4)	Agropine	White et al. (1985); Puddephat et al. (1996)
ATCC 15834	Agropine	Lippincott and Lippincott (1969); White and Nester (1980); Petit et al. (1983)
HRI	Agropine	Petit et al. (1983)
NCPPB1855	Agropine	Pomponi et al. (1983); Filetici et al. (1987)
NCPPB2659	Cucumopine	Combard et al. (1987); Filetici et al. (1987); Davioud et al. (1988)
K599	Cucumopine	Savka et al. (1992)
8196	Mannopine	Lippincott and Lippincott (1969); Petit et al. (1983); Koplow et al. (1984)
LMG63	Mannopine	Trypsteen et al. (1991)
LMG150	Mannopine	Trypsteen et al. (1991)
TR7	Mannopine	Lippincott and Lippincott (1969); Petit et al. (1983)
TR101	Mannopine	Lippincott and Lippincott (1969); Petit et al. (1983)
TR105	Mannopine	Maldonado-Mendoza et al. (1993)
MAFF30 1724	Mikimopine	Shiomi et al. (1987); Moriguchi et al. (2001)
MAFF-02-10266 A13	Mikimopine	Daimon et al. (1990)
R1000		Park and Facchini (2000)

nism by which the *rol* gene products cause the hairy root phenotype is still not completely understood.

In addition to the *rol* genes, the Ri T-DNA also contains genes for opine biosynthesis. Opines are unusual amino acid derivatives or sugar phosphodiesters that are produced in transformed tissues and secreted into the environment where they are catabolized preferentially by agrobacteria, thus giving the bacteria a selective advantage over other soil bacteria. Ri plasmids are classified into four groups based on the type of opine: agropine, mannopine, cucumopine, and mikimopine (Table 2.1), the latter being the stereo-isomer of cucumopine (Moriguchi et al. 2001). Sequence comparison of the Ri plasmid from strains 1724 and 2659 belonging to the mikimopine and the cucumopine types, respectively, has revealed that the opine catabolism mechanism is not closely related evolutionarily despite the fact that the corresponding opines are stereo-isomers (Moriguchi et al. 2001).

A. rhizogenes-mediated transformation of dicotyledonous plants has led to numerous applications in a broad array of research areas. Due to the high proliferation rate without the need for exogenous application of hormones, hairy root cultures are used to produce medicinal secondary compounds. The attractiveness of these hairy root cultures lies in their relative genetic and biochemical stability over long periods. This presents a clear advantage over cell suspension cultures that are more prone to erratic metabolic production because of their undifferentiated status. Moreover, the transformed roots often produce these secondary metabolites at higher levels than the non-transformed roots from which they are derived. In this context, it is noteworthy that hairy root cultures can be grown in bioreactors under closed conditions that minimize environmental risks, in contrast to field-cultivated transgenic plants. In spite of these advantages, few commercial successes have been obtained in the production of plant metabolites probably due to the fact that it is not cost-competitive. Other applications include artificial seed production and the study of responses towards chemical compounds (Mugnier 1997; Uozumi and Kobayashi 1997).

A. rhizogenes transformation is often used as a delivery system for foreign genes when a short-cut approach is preferred over the more elaborate and time-consuming *A. tumefaciens* transformation or when plants are recalcitrant to the latter bacteria. However, as a consequence of *rol* gene expression, regenerated plants are often characterized by morphological alterations, such as dwarfing, changed flowering, wrinkled leaves, and increased rooting and/or branching (Tepfer 1989). New techniques, such as the *rol*-type multi-autotransformation (MAT) vector system, eliminate these unwanted effects (Ebinuma et al. 1997a).

The *A. rhizogenes* transformation has been regarded as a particularly efficient tool to study different aspects of the interactions between roots and soil organisms. Bacteria belonging to the Rhizobiaceae family are capable of interacting symbiotically with roots of legume species (Schultze and Kondorosi 1998). This interaction is initiated through complex signal exchanges between both partners and results in the formation of nodules, new organs that house the differentiated bacteria during their nitrogen-fixing activity. *A. rhizogenes*-mediated transformation of legume roots has facilitated the introduction of gene constructs into these roots, resulting in so-called chimeric plants that carry wild-type upper parts and transgenic roots. Moreover, the nodulation capacities of these transformed roots are comparable with the wild-type situation, thus creating a rapid short-cut system for obtaining transgenic nodules (Stiller et al. 1997; Boisson-Dernier et al. 2001; Van de Velde et al. submitted). This strategy has been exploited to overexpress and suppress expression of genes involved in nodule development and functioning (Cheon et al. 1993). Promoter-trapping strategies that make use of hairy roots have been carried out to identify genes that form nodules (Martirani et al. 1999). Hairy roots have also been used to study the interactions between plant roots and nematodes (Kifle et al. 1999).

Hairy roots have also emerged as a valuable model system in studies on root-mediated phytoremediation (Shanks and Morgan 1999). This technology involves the use of plants to remove metals or organic contaminants from the soil.

The aim of this chapter is to present an overview of current techniques for different applications of the *A. rhizogenes* transformation system. Hand-on protocols are given that are used for a broad range of dicotyledonous plants. Pitfalls of these protocols will be thoroughly discussed.

2.2 Aspects In uencing *A. rhizogenes* Transformation Ef ciency

Although *A. rhizogenes* strains are capable of transforming a broad range of plant species, transformation frequencies can vary enormously between species and cultivars. Failures to transform certain species have been regularly reported (De Cleene and De Ley 1981; Tepfer 1990; Porter 1991). Therefore, it is essential to pinpoint certain important factors that influence the efficiency and reproducibility of transformation protocols. In broad terms, transformation efficiencies can be enhanced through handling of explants, choice of strains, and transformation parameters. A general scheme depicting the different steps of *A. rhizogenes* transformation and regeneration of plants is presented in Fig. 2.1. Inoculation of wounded explants with an *A. rhizogenes* inoculum is followed by a co-cultivation period on solid medium to allow transfer of T-DNA region(s) to plant cells. After this co-cultivation, agrobacteria are eliminated by incubating explants on medium containing appropriate antibiotics. After some time, adventitious roots start to grow from the infection site. A detailed protocol to obtain hairy roots is presented (Protocol 1), which was worked out for *Arabidopsis thaliana*, but has been used successfully for several plant species (Fig. 2.2).

2.2.1 Choice of *A. rhizogenes* Strain

One of the first steps in setting up an *A. rhizogenes* transformation experiment is choosing the appropriate *A. rhizogenes* strain. In addition to the type of opine that is synthesized by the transformed plant tissue, *A. rhizogenes* strains also differ in their T-DNA regions (Table 2.1). Agropine-type Ri plasmids possess two discrete T-DNAs, referred to as the left (T_L) and right (T_R) T-DNAs (Huffman et al. 1984; Jouanin 1984; De Paolis et al. 1985). During transformation, both T-DNA regions can integrate separately. The T_L region contains the *rol* genes, whereas genes involved in opine synthesis and auxin biosynthesis are found on the T_R region (reviewed in van der Salm et al. 1996). The mannopine-type, mikimopine-type, and cucumopine-type Ri plasmids have a single T-DNA region with *rol* genes and genes involved in opine synthesis (Koplow et al. 1984). For some plant species, the T_L T-DNA of the agropine-type plasmid is sufficient to induce hairy root formation; for others, the products encoded by the T_R-located auxin biosynthetic genes are essential for establishing hairy root syndromes (Cardarelli et al. 1985; White et al. 1985). The presence or absence of auxin synthesis genes could explain why the agropine Ri plasmid-containing strains are able to transform a wide range of plant species, whereas the mannopine strains have little hairy root-forming capacity on tissues with low endogenous auxin levels (Cardarelli et al. 1987; Filetici et al. 1987). Although, according to the literature, use of an agropine strain increases chances of obtaining hairy roots, the appropriate strain can be

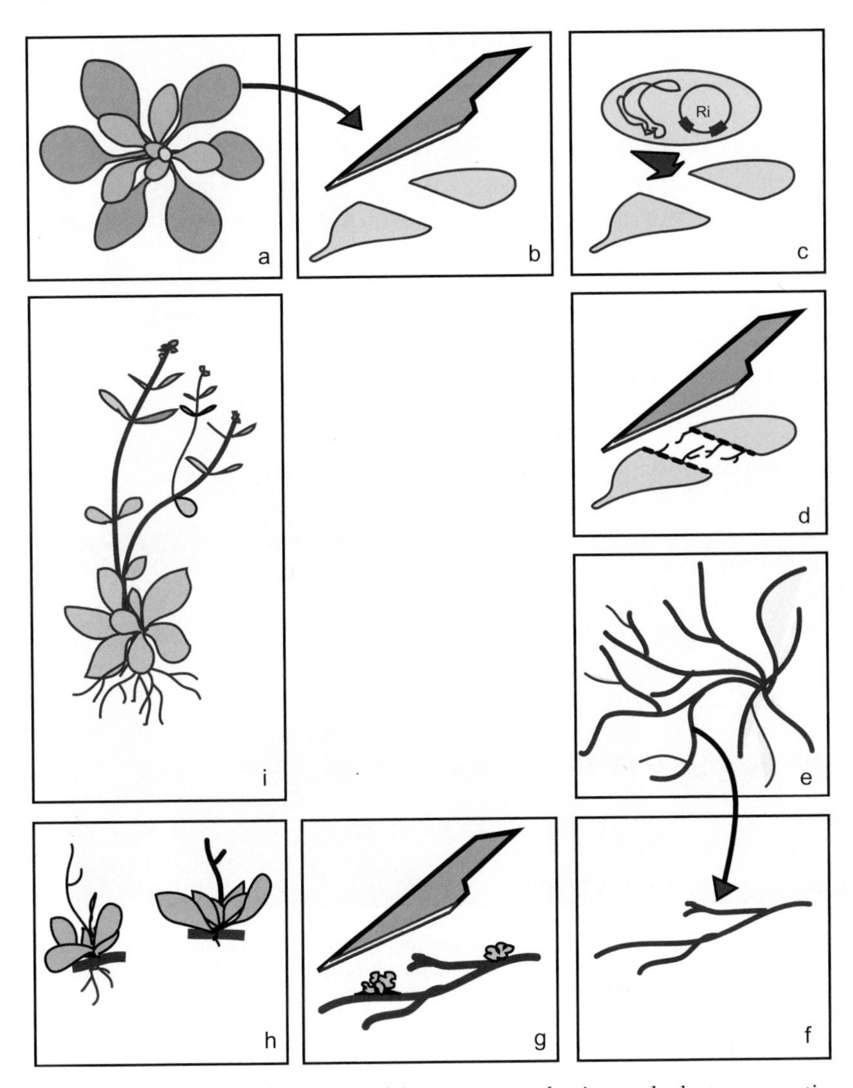

Fig. 2.1a–i. Schematic presentation of hairy root production and plant regeneration in *A. thaliana*. **a** Three-week-old *A. thaliana* plants; **b** dissection of leaf into two pieces and incubation on CIM; **c** co-cultivation with *A. rhizogenes*; **d** hairy roots emerging from the calli on cut sites of the leaves; **e** transformed hairy roots on H-EM; **f** incubation on H-CIM; **g** generation of green calli; **h** growing leaves from green calli on SIM; **i** shoots growing on EM. For medium description, see Table 2.2 page 38

selected with only one standard procedure, namely by testing a range of different *A. rhizogenes* strains. Differences in root growth, alkaloid production, and root morphology have been observed among various *A. rhizogenes* strains on the same plant species (Vanhala et al. 1995; Ionkova et al. 1997).

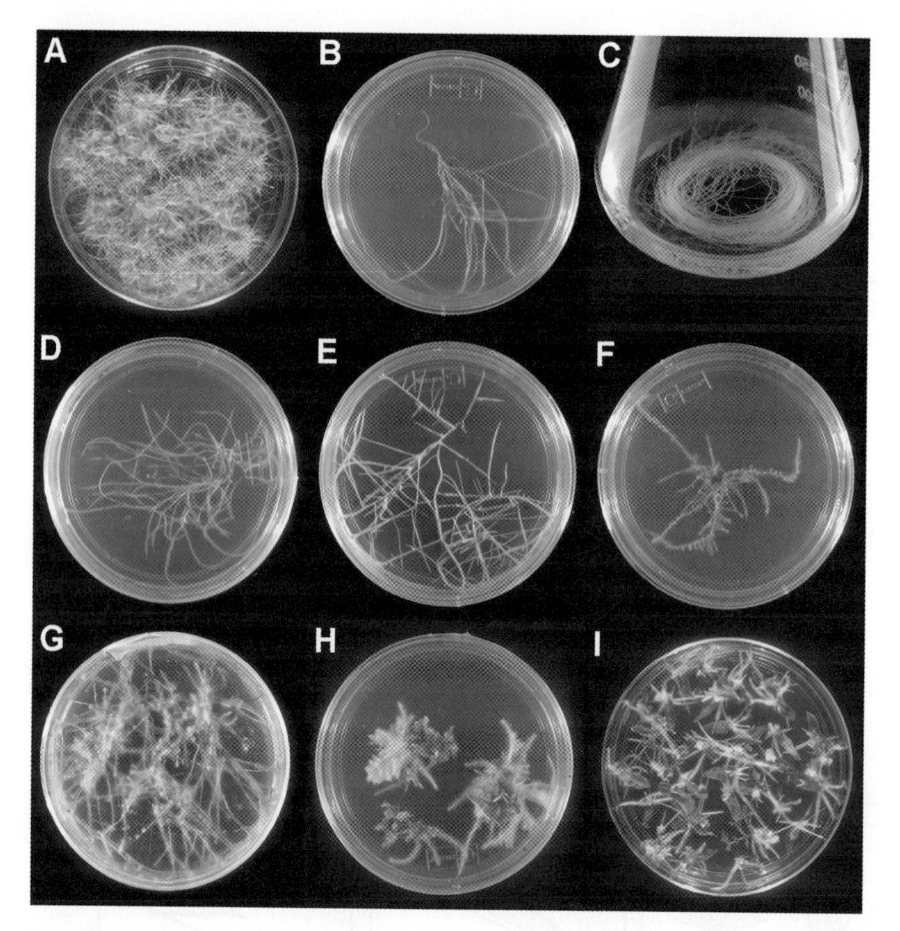

Fig. 2.2A–I. Hairy root growth of several plant species. *Agrobacterium rhizogenes* Ri 5834 and Protocol 1 were used for transformation. **A** *Arabidopsis*; **B** oilseed rape; **C** hairy root culture of oilseed rape in liquid H-EM (Table 2.2); **D** radish; **E** alfalfa; **F** mustard; **G** sugar beet; **H** tobacco; **I** tomato

2.2.2 Choice of Explant

Several types of explants, such as cotyledons, leaves, stems/hypocotyls, and root pieces have been used as starting material. The type of explant giving the highest transformation efficiency has to be determined experimentally for each species or cultivar. Young material, regardless of the tissue used, greatly enhances the transformation efficiency of hairy root formation (Rech et al. 1989; Trypsteen et al. 1991; Senior et al. 1995; Sanità di Toppi et al. 1997). As starting material, plants grown in vitro are preferred over those grown in greenhouses because this might increase transformation efficiencies (Tepfer 1984; Kifle et al. 1999).

2.2.3 Preparation of Bacterial Inoculum and Infection of Explants

The *A. rhizogenes* inoculum must always be prepared freshly, because otherwise the efficiency drops spectacularly. Cultures should also be grown at temperatures below 28–30°C because plasmid loss could occur at elevated temperatures (Sanità di Toppi et al. 1997).

Two main approaches can be used to prepare the inoculum. The first approach consists in growing a single colony on solid medium after which the freshly grown bacteria are used for the infection experiment. Either the bacteria are scraped off and directly applied onto the explants or the plate is washed with buffer, resulting in a bacterial suspension that is subsequently used for infection. Alternatively, an overnight liquid culture of the *A. rhizogenes* strain is used. In both cases, antibiotics should be omitted from the growth medium because of the negative effect on transformation efficiency. Inducers (acetosyringone or opines) of bacterial *vir* genes are often added to the *A. rhizogenes* growth medium to increase transformation frequency (Godwin et al. 1991; Fortin et al. 1992; Holford et al. 1992; Lipp João and Brown 1993). However, this positive effect greatly depends on the plant species. Solanaceae plants, for example, produce enough *vir* gene-inducing components and, therefore, require no extra inducing activity (Holford et al. 1992). Moreover, acetosyringone has been suggested to suppress virulence in some interactions between *A. rhizogenes* strains and plant species (Godwin et al. 1991; Vanhala et al. 1995). Opines have been shown to enhance the acetosyringone effect on *vir* gene induction (Veluthambi et al. 1988; Henzi et al. 2000). Therefore, pre-induction of the *A. rhizogenes* inoculum by a mixture of acetosyringone together with opines could enhance transformation efficiencies for certain hosts.

Routinely, explants are wounded before or simultaneously with the bacterial inoculation. Wounding triggers the secretion of *vir* gene-inducing compounds (Stachel et al. 1985) and can be done either with a sterile scalpel previously dipped into the bacterial inoculum or with a syringe, injecting small amounts of liquid *A. rhizogenes* inoculum. A now commonly used technique for *A. rhizogenes* infection consists of incubating wounded explants for a limited time (between 5 and 30 min) in a diluted bacterial suspension (ideally, $OD_{600} = 0.1–1.0$), followed by blotting off excess bacteria on sterile paper before the co-cultivation step (Karimi et al. 1999).

2.2.4 Co-cultivation

Several parameters need to be considered during the co-cultivation of agrobacteria and wounded explants. Generally, co-cultivation is performed on hormone-free in vitro plant growth media, such as MS (Murashige and Skoog 1962), B5 (Gamborg et al. 1968), or modified derivatives (Oksman-Caldentey 1991). To stimulate T-DNA transfer during co-cultivation, *vir* gene-inducing compounds can be added to the medium (200 μM acetosyringone). Although plant species/strain-specific observations have been made, a pH of 5.0–6.0 is optimal.

One of the most important factors is temperature. In several cases, elevated temperatures (higher than 20–22 °C) significantly lowered transformation frequencies (Stiller et al. 1997; Boisson-Dernier et al. 2001). A temperature optimum of 20–22°C has also been described for transient expression in *A. tumefaciens*-mediated transformation events (Dillen et al. 1997). Whether co-cultivation should be performed under dark or light (16 h/8 h) regimes has to be assessed experimentally because no consensus on this matter has been presented in the literature. The co-cultivation time is usually between 2 and 6 days. Too long an incubation time may result in overgrowth of the explants by agrobacteria. After explants have been co-cultivated, they are transferred to antibiotic-containing media to destroy the bacteria. The choice of antibiotics depends on the capacity to eliminate agrobacteria, but also on their potential hazardous effects on plant growth. For most agrobacterial strains cefotaxime (100–750 μg/ml) and/or carbenicillin (500–1000 μg/ml) can be used. In some cases, timentin (100–500 μg/ml), which is ticarcillin coupled to the β-lactamase inhibitor, clavulanic acid, was used (Cheng et al. 1998; Park and Facchini 2000). Carbenicillin and ticarcillin generally have no negative effect on plant growth, whereas cefotaxime has in certain cases (Yepes and Aldwinckle 1994; Shackelford and Chlan 1996; Hammerschlag et al. 1997; Ling et al. 1998). In a comparative study, timentin, carbenicillin, and cefotaxime were tested for their capacity to eliminate a hypervirulent *A. tumefaciens* strain from walnut somatic embryos (Tang et al. 2000). Timentin was the most effective because of the presence of clavulanic acid that inhibits β-lactamases produced by the agrobacteria. When selecting antibiotics to eliminate agrobacteria, the phytotoxicity should always be tested on the explants.

Another factor that might enhance transformation frequency is the addition of a non-oncogenic *A. tumefaciens* strain during infection. Co-inoculation of explants with both an *A. rhizogenes* strain and a cured *A. tumefaciens* strain results in a significantly higher number of hairy roots than with the *A. rhizogenes* strain alone (Torregrosa and Bouquet 1997; Kifle et al. 1999). An explanation could be that the *trans*-zeatin (*tsz*) gene, which is highly expressed in *A. tumefaciens* and whose product is involved in the synthesis of the cytokinin *trans*-zeatin, alters cytokinin levels in such a way that it has a beneficial effect on the transformation efficiency.

The use of plant-derived nurse cell layers during co-cultivation could also improve transformation rates (Fillatti et al. 1987; Christey and Sinclair 1992; Henzi et al. 2000). Usually tobacco or brassica cell suspension cultures are used and they probably stimulate *vir* gene expression during the co-cultivation period (Fillatti et al. 1987). Furthermore, addition of the auxin 2,4-dichlorophenoxyacetic acid (2,4-D) to the *A. rhizogenes* inoculum exerted a positive influence on hairy root formation of a difficult to transform Brassica cultivar (Puddephat et al. 2001).

2.3 Establishing the Transformed Nature of Hairy Roots

Hairy roots can grow from infection sites as well as from sites wounded accidentally by handling of the explants. In some cases, also growth of non-transformed roots has been observed (Ohara et al. 2000).

Hairy roots can be screened roughly based on their intrinsic characteristics that distinguish them from non-transformed roots. These characteristics include fast growth, heavy branching, and negative geotropism (Fig. 2.2). However, examples exist where transformed roots did not differ spectacularly from wild-type roots (Berthomieu and Jouanin 1992). More sound and safe tests are available to establish undoubtedly whether roots are the result of *A. rhizogenes* transformation.

The formation of hairy roots is the consequence of integration and expression of the *rol* genes, present on the T-DNA. To validate the presence of these *rol* genes in adventitious roots, DNA gel blot or polymerase chain reaction (PCR) analysis are performed, using *rol* gene-specific probes or primers, respectively. Contaminating *A. rhizogenes* DNA can be detected by using *vir* gene-specific probes or primers because these *vir* genes are not integrated into the plant cell genome (reviewed by Zupan et al. 2000).

Another frequently used approach to check whether roots are transformed is to check for the presence of opines. Upon integration of the Ri T-DNA into the plant cell genome, opine synthesis will be initiated through expression of the opine biosynthetic genes. The presence of opines in transformed roots is confirmed through high-voltage paper electrophoresis according to the method described by Petit et al. (1983). The agropine- and mannopine-type strains can be detected with silver nitrate staining (Petit et al. 1983), whereas for the cucumopine-type, the Pauly reagent is used as described by Savka et al. (1990). For agropine-type strains caution should be taken because the opine biosynthetic genes are located on the T_R T-DNA region, whereas *rol* genes are present on the T_L T-DNA region. Therefore, integration of the T_R and T_L T-DNA regions should always be checked through combined opine analysis (T_R T-DNA) and DNA gel blot analysis of *rol* genes (T_L T-DNA).

2.4 Co-transformation of Binary T-DNA

One of the major applications of *A. rhizogenes*-mediated transformation is the introduction of foreign genes into plants. This aim can be achieved by using either co-integrate or binary vectors. Co-integrate vectors carrying the foreign genes are integrated into the Ri T-DNA, implicating that the *rol* genes and genes of interest will never segregate. When the binary vector approach is used, the foreign genes are located on a T-DNA region that is separately mobilized in trans by the *vir* gene products of the Ri plasmid. Nowadays, the binary vector strategy is used commonly with different vectors being available on simple demand (www.cambia.org.au, www.pgreen.ac.uk). Meanwhile, the Gateway technology (Invitrogen, Carlsbad, CA) has revolutionized cloning into these vectors and several special Gateway binary vectors are accessible to the public (Karimi et al. 2002). This approach is particularly handy when transformed plants are wanted that are free of undesirable Ri-linked phenotypic effects associated with *rol* gene expression. Because the binary vector and Ri T-DNA have been inserted independently, both regions can segregate in subsequent generations. Although this independent segregation has been verified in several plant species (reviewed in Christey 2001), unsuccessful attempts have been reported as well (Senior et al. 1995; Christey 1997).

Usually, antibiotic resistance markers (mostly neomycin phosphotransferase II [NPTII]) are used to select co-transformed hairy roots. The efficiency of this selection method depends on the plant species. Although kanamycin selection has been used successfully (Boisson-Dernier et al. 2001), in other cases, for instance for certain species of the Brassicaceae family, low selection efficiency has been obtained because of growth prevention of transformed roots (Berthomieu and Jouanin 1992). Sometimes untransformed roots have been shown to be resistant against kanamycin (Hosoki and Kigo 1994), but this problem could be solved by using other aminoglycoside antibiotics, such as paromomycin (Park and Facchini 2000). The ineffectiveness of kanamycin and the efficacy of paromomycin in the selection procedure of transformed tissue have been reported before (Belny et al. 1997).

An important feature of *A. rhizogenes*-mediated transformation is that screening for transformed tissues can be done without the need for antibiotics, the use of which causes increased public concern. As an alternative for resistance markers, visual reporter genes are cloned on the binary T-DNA under control of a constitutive promoter (such as the cauliflower mosaic virus 35S promoter). Most commonly used is the β-glucuronidase (*gus*) reporter gene that catalyzes the hydrolysis of 5-bromo-4-chloro-3-indolyl-β-D-glucuronic acid, resulting in a blue precipitate (Jefferson et al. 1987). Unfortunately, analysis of GUS activity destroys hairy roots, making it impractical to monitor co-transformed hairy root growth in vivo. Other reporter systems, such as the firefly luciferase (*luc*) (reviewed by Greer and Szalay 2002) and green-fluorescent protein (*gfp*) genes (reviewed by Stewart 2001) are very suitable for non-destructive detection of co-transformed hairy roots (Stiller et al. 1997; Van de Velde et al., submitted). To visualize LUC proteins, a substrate has to be added that is oxidized by the luciferase, resulting in the release of photons of light (Bowie et al. 1973). The GFP marker has the advantage that it does not require any exogenous substrate or cofactor, except for oxygen, to be visualized (Chalfie et al. 1994). Upon excitation with blue light, GFP emits green-fluorescent light. Since the initial discovery of GFP in jellyfish, numerous GFP variants with improved spectral properties, fine-tuned for use in plants, have been created (reviewed by Stewart 2001). Using GFP as a reporter makes it also possible to screen large amounts of explants at the same time. To screen co-transformed hairy roots, the maize Sn protein has been proposed that activates the anthocyanin biosynthetic pathway, resulting in the production of a red pigment (Damiani et al. 1998). However, use of this protein as a general reporter is restricted.

Co-transformation frequencies can vary enormously and depend on the binary vector as well as on the specific combinations of plant species and strains used. Differential reporter gene expression between hairy root lines has been reported and can be attributed either to a different copy number of integrated T-DNAs or to position effects.

In addition to these positive selection and screening strategies, markers based on negative selection have been used (Stiller et al. 1997; Hashimoto et al. 1999). Ideally, these negative markers are used in a complementary manner with positive selection markers. This double strategy, integrating the negative selection marker into the Ri T-DNA, allows lines to be selected that only have the binary T-DNA carrying the positive marker.

2.5 Propagation of Hairy Root Lines in Liquid Cultures

Selected (co-)transformed hairy root lines can be propagated in liquid cultures for several purposes. One of the main applications of liquid hairy root cultures is the production of secondary metabolites that are secreted into the medium. The root lines can also be a useful tool to obtain satisfying amounts of starting material for the purification of DNA, RNA or proteins. As an example, a protocol is presented to propagate hairy roots in liquid culture (Protocol 3). However, as for all aspects of hairy root manipulation, no universal protocol exists to obtain healthy and fast growing hairy root cultures. Several parameters, such as the inoculum size, light conditions, and oxygen supply, have been analyzed in detail (Bhadra and Shanks 1995; Bourgaud et al. 1995; Kanokwaree and Doran 1997). In general, liquid hairy root cultures are grown under temperature/medium conditions similar to those for hairy roots on plates and need continuous shaking to stimulate aeration of the cultures. Light has been shown to have an inhibitory effect on hairy root growth for certain plant species (Bourgaud et al. 1995; Torregrossa and Bouquet 1997), but not for others (Karimi et al. 1999). Hairy roots can also be propagated on solid medium, using the same media and culture conditions as those for liquid cultures.

2.5.1 The Clonal Status of Hairy Roots

Hairy roots growing out of the infection site after *A. rhizogenes* transformation are presumably clonal, meaning that they are derived from a single transformed cell (Chilton et al. 1982). Variations at the level of growth rate, primary and secondary metabolism and phenotype between so-called clonal lines suggest that this is not always the case (Mano et al. 1989; Oksman-Caldentey et al. 1994; Sevón et al. 1995). The occasionally chimeric nature of hairy roots can be observed when expression patterns of reporter genes are monitored. Chimeric roots could result from integration of the binary T-DNA after the Ri T-DNA during co-transformation (Torregrosa and Bouquet 1997).

2.5.2 Stability of Long-Term Hairy Root Cultures

Maintaining hairy root lines on a long-term basis is useful for several applications. In general, hairy root lines are characterized by their genetic and biochemical stability (Maldonado-Mendoza et al. 1993). However, differences in alkaloid production in clonal lines over time have raised questions about the stability of the phenotype and the correlated gene expression in these lines. Loss of opine synthesis during long-term culture has been reported for hairy roots of several plant species (Tepfer 1984; Kamada et al. 1986). In addition, instability or even loss of expression of introduced genes into hairy roots has been reported (Guivarc'h et al. 1999). The latter observation has been linked to transcriptional and post-transcriptional silencing mechanisms. Complete inactivation or irregular expression of the transgenes in these cultures have been related to high copy numbers of inserted T-DNAs. Therefore, when long-term experiments are set up, this potential instability of hairy root lines should be taken into account, although it is surely not a general phenomenon.

2.6 Regeneration of Plants from Hairy Roots

Numerous plant species have been regenerated from transgenic hairy roots (see Christey 2001). Figure 2.1 gives an overview of the different steps leading to regenerated plants from hairy roots. In many of these cases, plants have been transformed with foreign genes, including agronomically useful traits. Until now, only experiments with dicotyledonous plants have been successful. Nevertheless, transient *gus* expression has recently been demonstrated in wheat cells using *A. rhizogenes* (Uzé et al. 2000). The *A. rhizogenes* transformation system is an attractive means to introduce foreign genes into plants because regeneration often occurs spontaneously and transformed plants can be obtained much faster than through *A. tumefaciens*-mediated transformation (Manners and Way 1989; Han et al. 1993). However, plants regenerated from hairy roots often display morphologically and physiologically altered phenotypes because of in planta expression of Ri T-DNA-associated gene expression that interferes with the hormone balance (Tepfer 1989). In some cases, these morphological changes are regarded as useful, for instance, to improve the ornamental qualities of certain horticultural species as well as to increase the rooting ability of woody species (reviewed in Christey 2001). For other applications, such as the introduction of agriculturally important traits into crop species, these morphogenic modifications are not desired. Examples exist in which the transgenically regenerated plants displayed no or minimal Ri-associated phenotypic alterations (Manners and Way 1989; Webb et al. 1990; Christey et al. 1994). Differential phenotypic responses of transformed plants are probably due to a combination of factors, such as intrinsic hormone characteristics of both plant and T-DNA and the plant-specific response towards *rol* gene expression. Several strategies can be used to obtain phenotypically normal plants, such as the use of a binary vector for introducing the foreign gene of interest followed by outcrossing of the Ri T-DNA or new methods, for instance the MAT transformation system that results in excision of the *rol* genes (see below).

The chimeric nature of some hairy roots could be a problem when such tissues are used for regeneration. A method to avoid the regeneration of chimeric plants might be to regenerate them from hairy root-derived protoplasts (Lee et al. 1995; Sevón et al. 1995).

Several procedures exist to obtain regenerated plants from hairy roots. Shoots can occur spontaneously or can be induced by transferring transformed tissues to hormone-containing media. The spontaneous shoot formation is particularly interesting because no callus step is involved, thus reducing chances of somaclonal variations. However, it is not a general characteristic of hairy roots and should, therefore, be evaluated experimentally for each plant–bacteria combination tested. For some plant species, spontaneous shooting has been observed when transformed roots are transferred from darkness to light (Noda et al. 1987; Petit et al. 1987). If shoots are not formed spontaneously, explants should be transferred to hormone-containing medium. Again, the quantity of growth regulators to be supplied has to be determined experimentally. Generally, a combination of auxin (usually indole-3-acetic acid or α-naphthaleneacetic acid) and cytokinine (6-benzylaminopurine) is used. After shoot formation, plantlets are put on root-inducing medium, adapted for the plant species under study. In many cases, this root growth medium does not contain any exogenously applied hormone.

An important consideration when trying to regenerate plants from hairy roots is the choice of the *A. rhizogenes* strain. Strain types that are competent in inducing hairy roots are not necessarily the best strains to use when shoot formation is the aim of the transformation experiment. As for the hairy root formation, several aspects need to be considered. For instance, lowering ethylene levels by using ethylene inhibitors could positively influence regeneration efficiency from hairy roots (Christey 1997).

2.7 The Multi-Auto Transformation (MAT) Vector System

As stated above, the use of *A. rhizogenes* strains to deliver genes into plants has several advantages over the *A. tumefaciens* transformation. One of the main benefits is that transformation can be screened without the use of antibiotic resistance markers, the insertion of which into transformed plants is currently one of the major public concerns regarding genetically modified organisms. However, a drawback is that regenerants can display *rol* gene-linked phenotypic alterations. Very recently, a strategy to eliminate the Ri T-DNA genes has been provided by the *rol*-type MAT vector system (reviewed by Ebinuma et al. 2001).

The MAT vector system exploits the morphological changes caused by the *Agrobacterium* oncogenes as selection markers and uses the recombinase/ recognition site (R/RS) recombination system of *Zygosaccharomyces rouxii* to remove the oncogenes before regeneration. Two systems exist: *rol* and isopentenyltransferase (*ipt*). The reader is also referred to http:// www.npaper.co.jp/. The first system uses the *rolA*, *rolB*, and *rolC* genes of *A. rhizogenes* 1724 strain as selection marker. The hairy root phenotype is used as a marker to screen for transformed plant cells. The second system integrates the *ipt* gene of *A. tumefaciens* as a selection marker, which causes an extremely shooty phenotype (ESP) and loss of apical dominance.

The system consists of a binary vector that not only contains the gene of interest between the T-DNA borders, but also a hit-and-run cassette. In this cassette, the *rolA*, *rolB*, and *rolC* genes of *A. rhizogenes* 1724 and the gene coding for the recombinase, under control of a CaMV 35S promoter, are present between two recognition sites. The binary vector is introduced into a non-oncogenic *A. tumefaciens* strain. After inoculation, the hairy root phenotype is a marker for transformed plant cells. In some cases, the cassette will be excised resulting in hairy roots with wild-type sectors. After a few weeks (to months), spontaneous regeneration may occur from the green segments of the hairy roots. Transformed wild-type shoots can be distinguished easily from the *rol* gene-containing shoots because the latter have wrinkled leaves, no apical dominance, and shortened internodes (Ebinuma et al. 1997b; Cui et al. 2000, 2001). Unfortunately, chimeric transgenic shoots that are difficult to discriminate are also observed. From these plants pure clones can be obtained by crossing and outsegregation (Ebinuma et al. 1997b). Due to early excision of the cassette, the efficiency of generating hairy roots is rather low (Ebinuma et al. 1997a, b; Cui et al. 2000). This problem can be solved in the future by using a construct in which the recombinase is under control of an inducible promoter that can be activated at the appropriate time (Sugita et al. 2000).

The system has been successfully used for tobacco and *Antirrhinum majus* (Cui et al. 2000, 2001). In tobacco, roots have to be transferred to shoot-inducing medium to grow shoots. Although the method seems very promising, more plant species, especially those that are recalcitrant to *A. tumefaciens* transformation, should be tested in the future.

2.8 Conclusions

Agrobacterium rhizogenes transformation is a reliable and fast method to obtain hairy roots that can be used in several research domains and for a number of applications. The regeneration of healthy plant species through *A. rhizogenes*-mediated transformation should profit from the advent of new techniques, such as the MAT vector system. In addition, a new successful *A. tumefaciens*-based transformation protocol has recently been described that dramatically improves the transformation efficiency (van der Fits et al. 2000). The technique involves the use of a ternary plasmid carrying a mutant *virG* gene that is expressed constitutively. Until now, the use of this constitutive *virG* construct has never been integrated into *A. rhizogenes*-mediated transformation experiments. Because the mechanism of T-DNA processing and transfer through the action of *vir* gene products is comparable for both *Agrobacterium* species, transformation efficiencies could benefit from the introduction of the ternary plasmid into the different *A. rhizogenes* strains used.

Protocol 1: Production of Transformed Hairy Roots

1. Plate 20–30 sterilized *A. thaliana* seeds on germination medium in Petri dishes (150 × 25 mm; Falcon, Becton Dickinson, Bedford, MA) and incubate for 3 weeks in a growth room at 22°C (16 h light/8 h dark photoperiod; Fig. 2.1a). For plant media, the reader is referred to Table 2.2.
2. Take leaves and cut them transversely in two halves with a sterile scalpel. Place the leaf halves upside-down on callus-inducing medium (Fig. 2.1b) and incubate for 4 days in the growth room. Note: to avoid tissue desiccation during cutting, float leaves on sterile water in Petri dishes.
3. Culture *A. rhizogenes* in 5 ml of liquid Luria broth (LB) medium (Invitrogen, Carlsbad, CA) without glucose (LB–) for 48 h at 28°C and shake at 200 rotations per minute. Make an infection solution by diluting the bacterial culture in liquid standard medium (SM) to a final $OD_{600} = 0.1$.
4. Incubate the leaf pieces in 20 ml of infection solution (Fig. 2.1c) in a Petri dish for 5 min and then blot them on sterile filter paper to remove excess liquid.
5. Transfer the leaf pieces to solidified SM medium; incubate for 3 days in the growth chamber.
6. Wash the leaf pieces three times with liquid SM medium containing 300 mg/l carbenicillin. Blot the explants (as in step 4) and place them on solidified SM medium containing 300 mg/l carbenicillin. Put 5 or 15 leaf pieces in 100- or 150-mm plates, respectively. Note (a): when the bacterium contains a T-DNA plasmid that carries a gene for antibiotic selection of transformed plant roots, include the appropriate antibiotic in the SM

Table 2.2. Plant media

Composition	GM	SM	CIM	H-EM	H-CIM	SIM	EM	RIM
MS[a]	1/2×	1×	1×	[b]	1×	1×	1×	1×
2-(*N*-Morpholino)ethanesulfonic acid (g/l)	–	0.5	0.5	–	0.5	0.5	0.5	0.5
Glucose (g/l)	–	20	20	–	–	–	–	–
Sucrose (g/l)	10	–	–	20	30	30	30	30
Phytagel (g/l)	3.5	3.5/–[c]	3.5	3.5/–[c]	3.5	3.5	3.5	3.5
pH (KOH)	5.8	5.8	5.8	6.2	5.8	5.8	5.8	5.8
2,4-Dichlorophenoxyacetic acid (mg/l)	–	–	0.2	–	–	–	–	–
Kinetin (mg/l)	–	–	0.2	–	4.3	–	–	–
α-Naphthaleneacetic acid (mg/l)	–	–	–	–	3.7	0.1	–	–
6(γ,γ-Dimethylallylamino)purine (mg/l)	–	–	–	–	–	2	–	–
6(γ,γ-Dimethylallylamino)purine-riboside (mg/l)	–	–	–	–	–	1	–	–
Indolebutyric acid (mg/l)	–	–	–	–	–	–	–	1
$AgNO_3$ (mg/l)	–	–	–	–	–	15	–	–

GM, germination medium; SM, standard medium; CIM, callus induction medium; H-EM, hairy root elongation medium ; H-CIM, hairy root callus-inducing medium; SIM, shoot-inducing medium; EM, elongation medium; RIM, root-inducing medium.
[a] Murashige and Skoog salts and vitamins (Murashige and Skoog 1962).
[b] Instead of MS, 1× De Greef and Jacobs salts (De Greef and Jacobs 1979).
[c] Solid/liquid.

medium. Note (b): when the bacterium contains a plasmid harboring a carbenicillin resistance gene, another antibiotic has to be used to eliminate the bacteria.

7. Incubate for 2 weeks on SM medium: transformed hairy roots start to emerge from small calli on the cut sites of the leaves (Fig. 2.1d).
8. Two weeks later, cut the hairy roots (1 cm long) and place them on hairy root elongation medium (H-EM; Fig. 2.1e) containing 200 mg/l carbenicillin (see note 6b; Fig. 2.2).
9. Subculture the transformed hairy roots every 2–3 weeks on H-EM medium without antibiotics. Hairy roots grow in the dark as well as in the light.

Protocol 2: Plant Regeneration from Hairy Roots

1. Incubate a few root pieces for 2 weeks on hairy root callus-inducing medium (H-CIM; Fig. 2.1f; Table 2.2; Van Sluys et al. 1987) containing naphthaleneacetic acid and kinetin (Table 2.2).
2. Transfer the root pieces with calli to shoot-inducing medium (SIM) for 4–6 weeks until shoots start to form on the green calli (Fig. 2.1g; Table 2.2).

3. Transfer these shoots to elongation medium (EM) for further growth. Most shoots change to small rosettes and form roots during incubation on EM medium (Fig. 2.1h, i; Table 2.2).
4. Incubate shoots that do not develop roots on root-inducing medium (RIM).

Protocol 3: Hairy Root Liquid Culture

1. Put 1–2 cm long pieces of hairy roots into a 250-ml Erlenmeyer flask containing 50 ml liquid H-EM (Fig. 2.2).
2. Place the flasks on a shaker (90 rpm) in the growth room at 22°C (16 h light/8 h dark) for 2–3 weeks until a mass of roots is obtained.

References

Belny M, Hérouart D, Thomasset B, David H, Jacquin-Dubreuil A, David A (1997) Transformation of *Papaver somniferum* cell suspension cultures with *sam1* from *A. thaliana* results in cell lines of different *S*-adenosyl-L-methionine synthetase activity. Physiol Plant 99:233–240

Berthomieu P, Jouanin L (1992) Transformation of rapid cycling cabbage (*Brassica oleracea* var. *capitata*) with *Agrobacterium rhizogenes*. Plant Cell Rep 11:334–338

Bhadra R, Shanks JV (1995) Statistical design of the effect of inoculum conditions on growth of hairy root cultures of *Catharanthus roseus*. Biotechnol Techn 9:681–686

Boisson-Dernier A, Chabaud M, Garcia F, Bécard G, Rosenberg C, Barker DG (2001) *Agrobacterium rhizogenes*-transformed roots of *Medicago truncatula* for the study of nitrogen-fixing and endomycorrhizal symbiotic associations. Mol Plant–Microbe Interact 14:695–700

Bourgaud F, Nguyen C, Guckert A (1995) Psoralea species: in vitro culture and production of furanocoumarins and other secondary metabolites. In: Bajaj YPS (ed) Medicinal and aromatic plants VIII. Biotechnology in agriculture and forestry, vol 33. Springer, Berlin Heidelberg New York, pp 388–411

Bowie LJ, Irwin R, Loken M, De Luca M, Brand L (1973) Excited-state proton transfer and the mechanism of action of firefly luciferase. Biochemistry 12:1852–1857

Cardarelli M, Spanò L, De Paolis A, Mauro ML, Nitali G, Costantino P (1985) Identification of the genetic locus responsible for non-polar root induction by *Agrobacterium rhizogenes* 1855. Plant Mol Biol 5:385–391

Cardarelli M, Spanò L, Mariotti D, Mauro ML, Van Sluys MA, Costantino P (1987) The role of auxin in hairy root induction. Mol Gen Genet 208:457–463

Chalfie M, Tu Y, Euskirchen G, Ward WW, Prasher DC (1994) Green fluorescent protein as a marker for gene expression. Science 263:802–805

Cheng Z-M, Schnurr JA, Kapaun JA (1998) Timentin as an alternative antibiotic for suppression of *Agrobacterium tumefaciens* in genetic transformation. Plant Cell Rep 17:646–649

Cheon C-I, Lee N-G, Siddique A-B M, Bal AK, Verma DPS (1993) Roles of plant homologs of Rab1p and Rab7p in the biogenesis of the peribacteroid membrane, a subcellular compartment formed *de novo* during root nodule symbiosis. EMBO J 12:4125–4135

Chilton M-D, Tepfer DA, Petit A, David C, Casse-Delbart F, Tempé J (1982) *Agrobacterium rhizogenes* inserts T-DNA into the genomes of the host plant root cells. Nature 295:432–434

Christey MC (1997) Transgenic crop plants using *Agrobacterium rhizogenes*-mediated transformation. In: Doran PM (ed) Hairy roots: culture and applications. Harwood Academic Publishers, Amsterdam, pp 99–111

Christey MC (2001) Use of Ri-mediated transformation of production of transgenic plants. In Vitro Cell Dev Biol Plant 37:687–700

Christey MC, Sinclair BK (1992) Regeneration of transgenic kale (*Brassica oleracea* var. *acephala*), rape (*B. napus*) and turnip (*B. campestris* var. *rapifera*) plants via *Agrobacterium rhizogenes* mediated transformation. Plant Sci 87:161–169

Christey MC, Sinclair BK, Braun RH (1994) Phenotype of transgenic *Brassica napus* and *B. oleracea* plants obtained from *Agrobacterium rhizogenes* mediated transformation. Abstract presented at the 8th International Congress on Plant tissue and cell culture, Florence, Italy

Combard A, Brevet J, Borowski D, Cam K, Tempé J (1987) Physical map of the T-DNA region of *Agrobacterium rhizogenes* strain NCPPB2659. Plasmid 18:70–75

Cui M, Takayanagi K, Kamada H, Nishimura S, Handa T (2000) Transformation of *Antirrhinum majus* L. by a *rol*-type multi-auto-transformation (MAT) vector system. Plant Sci 159:273–280

Cui M, Takayanagi K, Kamada H, Nishimura S, Handa T (2001) Efficient shoot regeneration from hairy roots of *Antirrhinum majus* L. transformed by the *rol* type MAT vector system. Plant Cell Rep 20:55–59

Daimon H, Fukami M, Mii M (1990) Hairy root formation in peanut by the wild type strains of *Agrobacterium rhizogenes*. Plant Tissue Cult Lett 7:31–34

Damiani F, Paolocci F, Consonni G, Crea F, Tonelli C, Arcioni S (1998) A maize anthocyanin transactivator induces pigmentation in hairy roots of dicotyledonous species. Plant Cell Rep 17: 339–344

Davioud E, Petit A, Tate ME, Ryder MH, Tempé J (1988) Cucumopine – a new T-DNA encoded opine in hairy root and crowngall. Phytochemistry 27:2429–2433

De Buck S, Jacobs A, Van Montagu M, Depicker A (1999) The DNA sequences of T-DNA junctions suggest that complex T-DNA loci are formed by a recombination process resembling T-DNA integration. Plant J 20:295–304

De Cleene M, De Ley J (1981) The host range of infectious hairy root. Bot Rev 47:147–194

De Greef W, Jacobs M (1979) In vitro culture of the sugarbeet: description of a cell line with high regeneration capacity. Plant Sci Lett 17:55–61

De Paolis A, Mauro ML, Pomponi M, Cardarelli M, Spanò L, Costantino P (1985) Localization of agropine-synthesizing functions in the T_R region of the root-inducing plasmid of *Agrobacterium rhizogenes* 1855. Plasmid 13:1–7

Dillen W, De Clercq J, Kapila J, Zambre M, Van Montagu M, Angenon G (1997) The effect of temperature on *Agrobacterium tumefaciens*-mediated gene transfer to plants. Plant J 12: 1459–1463

Ebinuma H, Sugita K, Matsunaga E, Yamakado M, Komamine A (1997a) Principle of MAT vector system. Plant Biotechnol 14:133–139

Ebinuma H, Sugita K, Matsunaga E, Yamakado M (1997b) Selection of marker-free transgenic plants using the isopentenyl transferase gene. Proc Natl Acad Sci USA 94:2117–2121

Ebinuma H, Sugita K, Matsunaga E, Endo S, Yamada K, Komamine A (2001) Systems for the removal of a selection marker and their combination with a positive marker. Plant Cell Rep 20:383–392

Filetici P, Spanò L, Costantino P (1987) Conserved regions in the T-DNA of different *Agrobacterium rhizogenes* root-inducing plasmids. Plant Mol Biol 9:19–26

Fillatti JJ, Sellmer J, McCown B, Haissig B, Comai L (1987) *Agrobacterium* mediated transformation and regeneration of *Populus*. Mol Gen Genet 206:192–199

Fortin C, Nester EW, Dion P (1992) Growth inhibition and loss of virulence in cultures of *Agrobacterium tumefaciens* treated with acetosyringone. J Bacteriol 174:5676–5685

Gamborg OL, Miller RA, Ojima K (1968) Nutrient requirements of suspension cultures of soybean root cells. Exp Cell Res 50:151–158

Godwin I, Todd G, Ford-Lloyd B, Newbury HJ (1991) The effects of acetosyringone and pH on *Agrobacterium*-mediated transformation vary according to plant species. Plant Cell Rep 9:671–675

Greer LF III, Szalay AA (2002) Imaging of light emission from the expression of luciferases in living cells and organisms: a review. Luminescence 17:43–74

Guivarc'h A, Boccara M, Prouteau M, Chriqui D (1999) Instability of phenotype and gene expression in long-term culture of carrot hairy root clones. Plant Cell Rep 19:43–50

Hammerschlag FA, Zimmerman RH, Yadava UL, Hunsucher S, Gercheva P (1997) Effects of antibiotics and exposure to an acidified medium on the elimination of *Agrobacterium tumefaciens* from apple leaf explants and on regeneration. J Am Soc Hortic Sci 122:758–763

Han KH, Keathley DE, Davis JM, Gordon MP (1993) Regeneration of a transgenic woody legume (*Robinia pseudoacacia* L, black locust) and morphological alterations induced by *Agrobacterium rhizogenes*-mediated transformation. Plant Sci 88:149–157

Hashimoto RY, Menck CFM, Van Sluys MA (1999) Negative selection driven by cytosine deaminase gene in *Lycopersicon esculentum* hairy roots. Plant Sci 141:175–181

Henzi MX, Christey MC, McNeil DL (2000) Factors that influence *Agrobacterium rhizogenes*-mediated transformation of broccoli (*Brassica oleracea* L. var. *italica*). Plant Cell Rep 19: 994–999

Holford P, Hernandez N, Newbury HJ (1992) Factors influencing the efficiency of T-DNA transfer during co-cultivation of *Antirrhinum majus* with *Agrobacterium tumefaciens*. Plant Cell Rep 11:196–199

Hosoki T, Kigo T (1994) Transformation of Brussels sprouts (*Brassica oleracea* var. *gemminifera* Zenk) by *Agrobacterium rhizogenes* harboring a reporter β-glucuronidase gene. J Jpn Soc Hortic Sci 63:589–592

Huffman GA, White FF, Gordon MP, Nester EW (1984) Hairy-root-inducing plasmid: physical map and homology to tumor-inducing plasmids. J Bacteriol 157:269–276

Ionkova I, Kartnig T, Alfermann W (1997) Cycloartane saponin production in hairy root cultures of *Astragalus mongholicus*. Phytochemistry 45:1597–1600

Jefferson RA, Kavanagh TA, Bevan MW (1987) GUS fusions: β-glucuronidase as a sensitive and versatile gene fusion marker in higher plants. EMBO J 6:3901–3907

Jouanin L (1984) Restriction map of an agropine-type Ri plasmid and its homologies with Ti plasmids. Plasmid 12:91–102

Kamada H, Okamura N, Satake M, Harada H, Shimomura K (1986) Alkaloid production of hairy root cultures in *Atropa belladonna*. Plant Cell Rep 5:239–242

Kanokwaree K, Doran PM (1997) The extent to which external oxygen transfer limits growth in shake flask culture of hairy roots. Biotechnol Bioeng 55:520–526

Karimi M, Van Montagu M, Gheysen G (1999) Hairy root production in *Arabidopsis thaliana*: cotransformation with a promoter-trap vector results in complex T-DNA integration patterns. Plant Cell Rep 19:133–142

Karimi M, Inzé D, Depicker A (2002) GATEWAY™ vectors for *Agrobacterium*-mediated plant transformation. Trends Plant Sci 7:193–195

Kifle S, Shao M, Jung C, Cai D (1999) An improved transformation protocol for studying gene expression in hairy roots of sugar beet (*Beta vulgaris* L.). Plant Cell Rep 18:514–519

Koplow J, Byrne MC, Jen G, Tempé J, Chilton M-D (1984) Physical map of the *Agrobacterium rhizogenes* strain 8196 virulence plasmid. Plasmid 11:17–27

Lee HS, Kim SW, Lee K-W, Eriksson T, Liu JR (1995) *Agrobacterium*-mediated transformation of ginseng (*Panax ginseng*) and mitotic stability of the inserted β-glucuronidase gene in regenerants from isolated protoplasts. Plant Cell Rep 14:545–549

Ling H-Q, Kriseleit D, Ganal MW (1998) Effect of ticarcillin/potassium clavulanate on callus growth and shoot regeneration in *Agrobacterium*-mediated transformation of tomato (*Lycopersicon esculentum* Mill.). Plant Cell Rep 17:843–847

Lipp João KH, Brown TA (1993) Enhanced transformation of tomato co-cultivated with *Agrobacterium tumefaciens* C58C1Rif^r::pGSFR1161 in the presence of acetosyringone. Plant Cell Rep 12:422–425

Lippincott BB, Lippincott JA (1969) Bacterial attachment to a specific wound site as an essential stage in tumor initiation by *Agrobacterium tumefaciens*. J Bacteriol 97:620–628

Maldonado-Mendoza IE, Ayora-Talavera T, Loyola-Vargas VM (1993) Establishment of hairy root cultures of *Datura stramonium*. Characterization and stability of tropane alkaloid production during long periods of subculturing. Plant Cell Tissue Organ Cult 33:321–329

Manners JM, Way H (1989) Efficient transformation with regeneration of the tropical pasture legume *Stylosanthes humilis* using *Agrobacterium rhizogenes* and a Ti plasmid-binary vector system. Plant Cell Rep 8:341–345

Mano Y, Ohkawa H, Yamada Y (1989) Production of tropane alkaloids by hairy root cultures of *Duboisia leichhardtii* transformed by *Agrobacterium rhizogenes*. Plant Sci 59:191–201

Martirani L, Stiller J, Mirabella R, Alfano F, Lamberti A, Radutoiu SE, Iaccarino M, Gresshoff PM, Chiurazzi M (1999) T-DNA tagging of nodulation- and root-related genes in *Lotus japonicus*: expression patterns and potential for promoter trapping and insertional mutagenesis. Mol Plant–Microbe Interact 12:275–284

Moore L, Warren G, Strobel G (1979) Involvement of a plasmid in the hairy root disease of plants caused by *Agrobacterium rhizogenes*. Plasmid 2:617–626

Moriguchi K, Maeda Y, Satou M, Hardayani NSN, Kataoka M, Tanaka N, Yoshida K (2001) The complete nucleotide sequence of a plant root-inducing (Ri) plasmid indicates its chimeric structure and evolutionary relationship between tumor-inducing (Ti) and symbiotic (Sym) plasmids in Rhizobiaceae. J Mol Biol 307:771–784

Mugnier J (1997) Mycorrhizal interactions and the effects of fungicides, nematicides and herbicides on hairy root cultures. In: Doran PM (ed) Hairy roots: culture and applications. Harwood Academic Publishers, Amsterdam, pp 123–131

Murashige T, Skoog F (1962) A revised medium for rapid growth and bio assays with tobacco tissue cultures. Physiol Plant 15:473–497

Noda T, Tanaka N, Mano Y, Nabeshima S, Ohkawa H, Matsui C (1987) Regeneration of horseradish hairy roots incited by *Agrobacterium rhizogenes* infection. Plant Cell Rep 6:283–286

Ohara A, Akasaka Y, Daimon H, Mii M (2000) Plant regeneration from hairy roots induced by infection with *Agrobacterium rhizogenes* in *Crotalaria juncea* L. Plant Cell Rep 19:563–568

Oksman-Caldentey K-M, Kivelä O, Hiltunen R (1991) Spontaneous shoot organogenesis and plant regeneration from hairy root cultures of *Hyoscyamus muticus*. Plant Sci 78:129–136

Oksman-Caldentey K-M, Sevón N, Vanhala L, Hiltunen R (1994) Effect of nitrogen and sucrose on the primary and secondary metabolism of transformed root cultures of *Hyoscyamus muticus*. Plant Cell Tissue Org Cult 38:263–272

Park S-U, Facchini PJ (2000) *Agrobacterium rhizogenes*-mediated transformation of opium poppy, *Papaver somniferum* L, California poppy, *Eschscholzia californica* Cham, root cultures. J Exp Bot 51:1005–1016

Petit A, David C, Dahl GA, Ellis JG, Guyon P, Casse-Delbart F, Tempé J (1983) Further extension of the opine concept: plasmids in *Agrobacterium rhizogenes* cooperate for opine degradation. Mol Gen Genet 190:204–214

Petit A, Stougaard J, Kühle A, Marcker KA, Tempé J (1987) Transformation and regeneration of the legume *Lotus corniculatus*: a system for molecular studies of symbiotic nitrogen fixation. Mol Gen Genet 207:245–250

Pomponi M, Spanò L, Sabbadini MG, Costantino P (1983) Restriction endonuclease mapping of the root-inducing plasmid of *Agrobacterium rhizogenes* 1855. Plasmid 10:119–129

Porter JR (1991) Host range and implications of plant infection by *Agrobacterium rhizogenes*. Crit Rev Plant Sci 10:387–421

Puddephat IJ, Riggs TJ, Fenning TM (1996) Transformation of *Brassica oleracea* L.: a critical review. Mol Breeding 2:185–210

Puddephat IJ, Robinson HT, Fenning TM, Barbara DJ, Morton A, Pink DAC (2001) Recovery of phenotypically normal transgenic plants of *Brassica oleracea* upon *Agrobacterium rhizogenes*-mediated co-transformation and selection of transformed hairy roots by GUS assay. Mol Breed 7:229–242

Rech EL, Golds TJ, Husnain T, Vainstein MH, Jones B, Hammatt N, Mulligan BJ, Davey MR (1989) Expression of a chimaeric kanamycin resistance gene introduced into the wild soybean *Glycine canescens* using a cointegrate Ri plasmid vector. Plant Cell Rep 8:33–36

Riker A, Banfield W, Wright W, Keitt G, Sagen H (1930) Studies on infectious hairy root of nursery apple trees. J Agric Res 41:887–912

Sanità di Toppi L, Pecchioni N, Durante M (1997) *Cucurbita pepo* L. can be transformed by *Agrobacterium rhizogenes*. Plant Cell Tissue Organ Cult 51:89–93

Savka MA, Ravillion B, Noel GR, Farrand SK (1990) Induction of hairy roots on cultivated soybean genotypes and their use to propagate the soybean cyst nematode. Phytopathology 80: 503–508

Savka MA, Liu L, Farrand SK, Berg RH, Dawson JO (1992) Induction of hairy roots or pseudoactinorhizae on *Alnus glutinosa, A. acuminata* and *Elaeagnus angustifolia* by *Agrobacterium rhizogenes*. Acta Oecol 13:423–431

Schultze M, Kondorosi A (1998) Regulation of symbiotic root nodule development. Annu Rev Genet 32:33–57

Senior I, Holford P, Cooley RN, Newbury HJ (1995) Transformation of *Antirrhinum majus* using *Agrobacterium rhizogenes*. J Exp Bot 46:1233–1239

Sevón N, Oksman-Caldentey K-M, Hiltunen R (1995) Efficient plant regeneration from hairy root-derived protoplasts of *Hyoscyamus muticus*. Plant Cell Rep 14:738–742

Shackelford NJ, Chlan CA (1996) Identification of antibiotics that are effective in eliminating *Agrobacterium tumefaciens*. Plant Mol Biol Rep 14:50–57

Shanks JV, Morgan J (1999) Plant "hairy root" culture. Curr Opin Biotechnol 10:151–155

Shiomi T, Shirakata T, Takeuchi A, Oizumi T, Uematsu S (1987) Hairy root of melon caused by *Agrobacterium rhizogenes* biovar 1. Ann Phytopath Soc Jpn 53:454–459

Stachel SE, Messens E, Van Montagu M, Zambryski P (1985) Identification of the signal molecules produced by wounded plant cells that activate T-DNA transfer in *Agrobacterium tumefaciens*. Nature 318:624–629

Stewart CN Jr (2001) The utility of green fluorescent protein in transgenic plants. Plant Cell Rep 20:376–382

Stiller J, Martirani L, Tuppale S, Chian R-J, Chiurazzi M, Gresshoff PM (1997) High frequency transformation and regeneration of transgenic plants in the model legume *Lotus japonicus*. J Exp Bot 48:1357–1365

Sugita K, Kasahara T, Matsunaga E, Ebinuma H (2000) A transformation vector for the production of marker-free transgenic plants containing a single copy transgene at high frequency. Plant J 22:461–469

Tang H, Ren Z, Krczal G (2000) An evaluation of antibiotics for the elimination of *Agrobacterium tumefaciens* from walnut somatic embryos and for the effects on the proliferation of somatic embryos and regeneration of transgenic plants. Plant Cell Rep 19:881–887

Tepfer D (1984) Transformation of several species of higher plants by *Agrobacterium rhizogenes*: sexual transmission of the transformed genotype and phenotype. Cell 37:959–967

Tepfer D (1989) Ri T-DNA from *Agrobacterium rhizogenes*: a source of genes having applications in rhizosphere biology and plant development, ecology, and evolution. In: Kosuge T, Nester EW (eds) Plant-microbe interactions: molecular and genetic perspectives, vol 3. McGraw-Hill, New York, pp 294–342

Tepfer D (1990) Genetic transformation using *Agrobacterium rhizogenes*. Physiol Plant 79: 140–146

Torregrosa L, Bouquet A (1997) *Agrobacterium rhizogenes* and *A. tumefaciens* co-transformation to obtain grapevine hairy roots producing the coat protein of grapevine chrome mosaic nepovirus. Plant Cell Tissue Organ Cult 49:53–62

Trypsteen M, Van Lijsebettens M, Van Severen R, Van Montagu M (1991) *Agrobacterium rhizogenes*-mediated transformation of *Echinacea purpurea*. Plant Cell Rep 10:85–89

Uozumi N, Kobayashi T (1997) Artificial seed production through hairy root regeneration. In: Doran PM (ed) Hairy roots: culture and applications. Harwood Academic Publishers, Amsterdam, pp 113–121

Uzé M, Potrykus I, Sautter C (2000) Factors influencing T-DNA transfer from *Agrobacterium* to precultured immature wheat embryos (*Triticum aestivum* L.). Cereal Res Commun 28:17–23

van der Fits L, Deakin EA, Hoge JHC, Memelink J (2000) The ternary transformation system: constitutive *virG* on a compatible plasmid dramatically increases *Agrobacterium*-mediated plant transformation. Plant Mol Biol 43:495–502

van der Salm TPM, Hänisch ten Cate CH, Dons HJM (1996) Prospects for applications of *rol* genes for crop improvement. Plant Mol Biol Rep 14:207–228

Van Sluys MA, Tempé J, Fedoroff N (1987) Studies on the introduction and mobility of the maize *Activator* element in *Arabidopsis thaliana* and *Daucus carota*. EMBO J 6:3881–3889

Vanhala L, Hiltunen R, Oksman-Caldentey K-M (1995) Virulence of different *Agrobacterium* strains on hairy root formation of *Hyoscyamus muticus*. Plant Cell Rep 14:236–240

Veluthambi K, Ream W, Gelvin SB (1988) Virulence genes, borders, and overdrive generate single-stranded T-DNA molecules from the A6 Ti plasmid of *Agrobacterium tumefaciens*. J Bacteriol 170:1523–1532

Webb KJ, Jones S, Robbins MP, Minchin FR (1990) Characterization of transgenic root cultures of *Trifolium repens*, *Trifolium pratense* and *Lotus corniculatus* and transgenic plants of *Lotus corniculatus*. Plant Sci 70:243–254

White FF, Nester EW (1980) Hairy root: plasmid encodes virulence traits in *Agrobacterium rhizogenes*. J Bacteriol 141:1134–1141

White FF, Ghidossi G, Gordon MP, Nester EW (1982) Tumor induction by *Agrobacterium rhizogenes* involves the transfer of plasmid DNA to the plant genome. Proc Natl Acad Sci USA 79:3193–3197

White FF, Taylor BH, Huffman GA, Gordon MP, Nester EW (1985) Molecular and genetic analysis of the transferred DNA regions of the root-inducing plasmid of *Agrobacterium rhizogenes*. J Bacteriol 164:33–44

Willmitzer L, Sanchez-Serrano J, Buschfeld E, Schell J (1982) DNA from *Agrobacterium rhizogenes* is transferred to and expressed in axenic hairy root plant tissues. Mol Gen Genet 186:16–22

Yepes LM, Aldwinckle HS (1994) Factors that affect leaf regeneration efficiency in apple, and effect of antibiotics in morphogenesis. Plant Cell Tissue Organ Cult 37:257–269

Zupan J, Muth TR, Draper O, Zambryski P (2000) The transfer of DNA from *Agrobacterium tumefaciens* into plants: a feast of fundamental insights. Plant J 23:11–28

3 Transformation of *Petunia hybrida* by the *Agrobacterium* Suspension Drop Method

S.J. WYLIE, D. TJOKROKUSUMO, and J.A. MCCOMB

3.1 Introduction

Plant transformation is a key methodology that has allowed transfer and expression of novel genes for the improvement of economically important plant species as well as enquiry into deeper questions about the function of plant genes. For many plant species, stable transformation remains difficult or impossible. Where it is possible, there is usually a need for expensive resources such as laminar flow hoods, controlled environment growth rooms and highly skilled practitioners. In addition, there are often problems related to combining efficient plant regeneration with gene transfer as transfer techniques are carried out in undifferentiated cell cultures. Low transformation efficiency, instability of transgene expression, somaclonal variation and inability to regenerate whole plants are common problems.

Gene transfer technologies for plants can be broadly divided into vector-mediated methods and direct gene methods (Hooykaas and Schilperoort 1992; Christou 1996). Here, we look at transformation mediated by the plant pathogenic bacterial vector *Agrobacterium tumefaciens*. *Agrobacterium*-mediated transformation occurs when wounded plant tissue is exposed to *Agrobacterium* cells containing a plasmid with the gene of interest and a selectable marker gene located within the transferred DNA (T-DNA) region. The bacterium is stimulated to expresses virulence (Vir) proteins by exposure to phenolic compounds exuded from wounded cells or by addition of acetosyringone to a cell culture (Hoekema et al. 1983). The Vir proteins are responsible for the excision, transfer and integration of the T-DNA into the plant genome.

The T-DNA is transferred to only a small proportion of cells. In in-vitro transformation/regeneration systems, the transformed cells are given a selective advantage over the relatively large number of non-transformed cells by exposure to a selection agent. The transgenic cells carry a gene that encodes a protein that either inactivates the selection agent, thereby allowing the transformed cells to thrive while the untransformed cells die or, in the case of positive selection markers (Joersbo and Okkels 1996), allows the transformed cells to utilise a carbon source that is unavailable to the wild-type cells.

In-vitro methods of *Agrobacterium*-mediated transformation require that the explants, tissue slices, callus or protoplasts be carefully prepared, often under a microscope, grown under sterile, climate-controlled conditions, with

Molecular Methods of Plant Analysis, Vol. 23
Genetic Transformation of Plants

frequent subculture. The technique results in the isolation of a low number of transformed cells, and protocols for organogenesis or embryogenesis from these cells are then required. The whole process requires considerable infrastructure and is labour-intensive.

Seeds, young seedlings and mature plants of the model plant *Arabidopsis thaliana* have been transformed with *Agrobacterium* in planta, thus avoiding the need for in-vitro technology (Feldmann and Marks 1987; Feldmann 1992; Bechtold et al. 1993; Chang et al. 1994; Katavic et al. 1994; Clough and Bent 1998). In the original whole plant technique, plants were submerged in *Agrobacterium* in a nutrient medium supplemented with sucrose and plant hormones, then a vacuum was applied (Bechtold et al. 1993). Later, the floral dip method was reported where only inflorescences were dipped in a solution containing *Agrobacterium*, sucrose and a surfactant, then vacuum-treated (Clough and Bent 1998). Unfortunately, this method has not been applied routinely to in-planta transformation of other species. Similar transformation approaches for the model legume *Medicago truncatula* (barrel medic; Trieu et al. 2000), *Glycine max* (soybean; Hu and Wang 1999) and *Brassica rapa* ssp *chinensis* plants (Pakchoi; Qing et al. 2000) are reported, although none are used routinely.

Several methods of co-cultivation of pollen with *Agrobacterium tumefaciens* and *A. rhizogenes* were tested for transformation of *Nicotiana langsdorf i* without success (Sanford and Skubik 1986). *Agrobacterium*-mediated transformation using pollen or pollen tubes have been reported, although in most cases DNA integration in the progeny has been unstable (Hess et al. 1990; Langridge et al. 1992; Zheng et al. 1994). Techniques for direct transformation of microspores and pollen by particle bombardment have been described for monocots and dicots (e.g. Ramaiah and Skinner 1997; Mentewab et al. 1999; Fernando et al. 2000).

Non-*Agrobacterium* based methods for germline transformation are reported for soybean, including injection of DNA directly to the ovary, and introduction of total genomic DNA via the pollen tube pathway (Hu and Wang 1999). A drop of DNA placed directly onto the cut end of rice florets was reported to result in transformation of up to 20% of the seedlings (Luo and Wu 1989). Mixing pollen with plasmid DNA in-vitro before pollination resulted in no transformation of maize embryos (Booy et al. 1989).

All these methods avoid the use of a selection agent immediately after transformation; instead the progeny seed are screened on the selection agent as they germinate, or a selection agent is not used, but the modified trait is detected. Most transformed progeny are non-chimeric and there is not the somaclonal variation associated with tissue culture. Apart from the protocols that describe pollen transformation or ovule injection, the specific cell type that is transformed is unknown, but it is likely that it is the gametophyte-progenitor tissues, mature gametophytes, or recently fertilised embryos (Clough and Bent 1998). In floral dip methods, the transformation event probably occurs late in flower development (Desfeux et al. 2000).

We investigated two methods of introducing the selectable marker gene *bar* and the reporter gene *gus* to the germline of *Petunia hybrida*; vacuum infiltrating pollen with *Agrobacterium tumefaciens*, and application of *Agrobacterium* cells directly to the flower stigma at pollination. We found that transformation was achieved with similar frequencies by each method (Tjokrokusumo et al. 2000), indicating that the vacuum treatment did not play a significant role in the transformation process. The *Agrobacterium* suspension drop method is simple to apply and requires no special equipment such as a vacuum chamber and pump. Here, we describe the *Agrobacterium* suspension drop method for transformation of *Petunia hybrida*.

3.2 Transformation

Agrobacterium tumefaciens strain AgL0 (Lazo et al. 1991) was cultured in GYPC medium (2 g/l glucose, 2 g/l yeast extract, 15 mM potassium phosphate buffer (pH 7.0), 2 g/l casein hydrolysate) with appropriate antibiotics until an OD_{600} of 0.80 was reached. This *Agrobacterium* strain carries a binary plasmid pCGP1258 which contains the *uid A* (*gus*) gene encoding β-glucuronidase, and a herbicide resistance gene, *bar*, encoding phosphinothricin acetyl-transferase, both controlled by CaMV 35S constitutive promoters. The *uid A* gene is not expressed in bacteria because of the deletion of the bacterial ribosomal binding site (Jansen and Gardner 1989). One millilitre of *A. tumefaciens* suspension was placed in a sterile 1.5-ml centrifuge tube and centrifuged at 18,000 ×g for 3 min. The pellet was resuspended in 1 ml of pollen germination medium (20% sucrose with 100 ppm H_3BO_3 and 300 ppm $CaCl_2.2H_2O$) and again centrifuged at 18,000 ×g for 3 min and the supernatant discarded. The bacterial pellet was resuspended in enough pollen germination medium to make a thick solution. A drop of the bacterial solution was applied directly to the stigmas of *Petunia hybrida* cultivar Peach Ice emasculated flowers, followed by pollination. Pollen was collected from flowers with recently dehisced anthers and applied to treated stigmas with a small paintbrush. Isolation bags were kept on the flowers for 2 weeks after pollination. In control plants, a mock treatment of pollen germination medium without bacteria was applied to the stigma before pollination.

3.3 Analysis of Transformants

3.3.1 Screening *Petunia* Seedlings for Herbicide Resistance

Petunia seeds from controls and treated plants were surface-sterilised with 70% ethanol for 30 s, given two rinses in sterile water, then 5 min in 0.2%

sodium hypochlorite followed by four rinses in sterile water. The seeds were germinated at 25°C, 16 h light (36 μmol m^{-2} s^{-1}), on Murashige and Skoog (1962) nutrient salts medium with 20 g/l sucrose, 8 g/l agar, and 3 mg/l of glufosinate in the form of technical grade Basta in 9-cm Petri dishes. After 4 weeks, seedlings susceptible to Basta were either dead, yellow, or had developed only to the cotyledon stage. Putatively transformed resistant plants were green and had true primary leaves.

Amongst treated plants there were 0–20% resistant seedlings per capsule. Discrimination from the controls was not absolute as at 4 weeks, 1% of control seedlings survived on selection medium, although they died by 8 weeks. It was not possible to screen all seedlings for 8 weeks on the Basta medium as after 8 weeks resistant seedlings were slow to resume growth when returned to medium without Basta. Consequently, it is recommended that selection of resistant plants is done at 4 weeks, recognising that the genotypes selected at this time may include a low number of non-transformed plant escapes. It is possible that better discrimination of resistant plants might be obtained by growing all seedlings under glasshouse conditions and spot treating leaves with a 50 mg/l Basta solution.

3.3.2 Transmission of Basta Resistance Phenotype to T_2 Progeny

A lower than expected frequency of Basta-resistant T_2 progeny was observed from both a backcross to a control plant, and a cross between two transformed T_1 plants. Previous studies have shown that T_1 transformants generated by similar methods are typically hemizygous, carrying T-DNA at only one of two alleles of a given locus (Feldmann 1991; Bechtold et al. 1993). The expected proportions of resistant progeny, where one or both of the resistant parents are heterozygous for the transgene, are 50 and 75% respectively. It is possible that expression of the gene was low, or the transgene reduced gamete or seed fitness.

3.3.3 β-Glucuronidase Assay

Expression of β-glucuronidase was evident in 14% of T1 plants and 0–6% of T2 plants. Where there was strong expression in leaves, it was also high in anthers (both immature and mature), the internal tissues of pistils and low in sepals. The leaves of the control plants showed no expression, but a weak blue stain was seen in mature, but not in immature anthers, and there was no stain in pistils. There was no blue staining in the roots of either transformed plants or control plants. Variation in the level of β-glucuronidase expression between transformed plants suggests different numbers of transgene copies (Hobbs et al. 1993), different methylation patterns of the chromosomal integration region (Prols and Meyer 1992), or differences in penetration of the staining substrate or transcriptional or post-transcriptional silencing of the transgene (Martin et al. 1992). Similar reasons have been advanced to explain the non-expression of the *gus* gene in transgenic plants of other species (Ohta 1986; Kilby et al. 1992; Prols and Meyer 1992; Ottaviani et al. 1993; Meyer et al. 1994; Meyer 1995; Senior and Dale 1996).

3.3.4 DNA Analysis

The presence of transgenes was detected by PCR in 60–85% of the putatively transformed plants selected on Basta medium. As mentioned above, it is possible that a low number of non-transformed individuals could pass the 4-week screening on the selection medium, but it is also possible that the negative plants were due to incomplete insertion or unstable integration into the genome (Hess et al. 1990; Langridge et al. 1992).

To check that the technique resulted in independent transformation events, Southern blot analysis was performed (Southern 1975) on digested genomic DNA from some of the plants. Different banding patterns differentiated discrete transformation events. The plants analysed had two or three copies of the transgene. Plants in the T_2 generation were similarly analysed and showed that the transgenes were stably inherited.

3.4 Conclusion

The method described here, where pollen and *Agrobacterium* are combined on the receptive stigmas of emasculated flowers, is suitable for gene transfer to *Petunia hybrida*. The results of Southern blotting and PCR analysis confirmed that integration of transgenes was stable and that they were inherited. It is probable that this method, or modifications of it, will be suitable for generating transgenic plants of other species of the Solanaceae, and possibly those of other families, particularly those with high numbers of seeds per fruit, self-incompatible species, or tree species for which other techniques such as vacuum infiltration would be impossible. We have achieved some promising early results with transformation of *Zea mays*, but further experimental work needs to be done with this species.

References

Bechtold N, Ellis J, Pelletier G (1993) *In planta Agrobacterium*-mediated gene transfer by infiltration of adult *Arabidopsis thaliana* plants. CR Acad Sci Paris Life Sci 316:1194–1199

Booy G, Krens FA, Huizing HJ (1989) Attempted pollen-mediated transformation of maize. J Plant Physiol 135:319–324

Chang SS, Park SK, Kim BC, Kang BJ, Kim DU, Nam HG (1994) Stable genetic transformation of *Arabidopsis thaliana* by *Agrobacterium* inoculation in planta. Plant J 5:551–558
Christou P (1996) Transformation technology. Trend Plant Sci 1:423–431
Clough SJ, Bent AF (1998) Floral dip: a simplified method for *Agrobacterium*-mediated transformation of *Arabidopsis thaliana*. Plant J 16:735–743
Dellaporta S, Wood J, Hick J (1983) A plant DNA minipreparation; version II. Plant Mol Biol Rep 1:19–21
Desfeux C, Clough SJ, Bent AF (2000) Female reproductive tissues are the primary target of *Agrobacterium*-mediated transformation by the *Arabidopsis* floral-dip method. Plant Physiol 123:895–904
Feldmann KA (1991) T-DNA insertion mutagenesis in *Arabidopsis*: mutational spectrum. Plant J 1:71–83
Feldmann K (1992) T-DNA insertion mutagenesis in *Arabidopsis*: seed infection transformation. In: Koncz C, Chua N-H, Schell J (eds) Methods in *Arabidopsis* research. World Scientific, Singapore, pp 274–289
Feldmann K, Marks M (1987) *Agrobacterium*-mediated transformation of germinating seeds of *Arabidopsis thaliana*: a non-tissue culture approach. Mol Gen Genet 208:1–9
Fernando DD, Owens JN, Misra S (2000) Transient gene expression in pine pollen tubes following particle bombardment. Plant Cell Rep 19:224–228
Hess D, Dressler K, Nimmrichter R (1990) Transformation experiments by pipetting *Agrobacterium* into the spikelets of wheat (*Triticum aestivum* L.). Plant Sci 72:233–244
Hobbs SLA, Warkentin TD, Delong CMO (1993) Transgene copy number can be positively or negatively associated with transgene expression. Plant Mol Biol 21:17–26
Hoekema A, Hirsch PR, Hooykaas PJJ, Schilperoort RA (1983) A binary plant vector strategy based on separation of *vir*- and T-region of the *Agrobacterium tumefaciens* Ti-plasmid. Nature 303: 179–180
Hooykaas PJJ, Schilperoort RA (1992) *Agrobacterium* and plant genetic engineering. Plant Mol Biol 13:327–336
Hu CY, Wang LZ (1999) *In planta* soybean transformation technologies developed in China: procedure, confirmation and field performance. In Vitro Cell Devel Biol-Plant 35:417–420
Janssen BJ, Gardner RC (1989) Localized transient expression of GUS in leaf disks following cocultivation with *Agrobacterium*. Plant Mol Biol 14:61–72
Jefferson RA (1987) Assaying chimeric genes in plants: the *gus* gene fusion system. Plant Mol Biol Rep 5:387–405
Joersbo M, Okkels FT (1996) A novel principle for selection of transgenic plant cells – positive selection. Plant Cell Rep 16:219–221
Katavic V, Haughn GW, Reed D, Martin M, Kunst L (1994) *In planta* transformation of *Arabidopsis thaliana*. Mol Gen Genet 245:363–370
Kilby NJ, Leyser HMO, Furner IJ (1992) Promoter methylation and progressive transgene inactivation in *Arabidopsis*. Plant Mol Biol 20:103–112
Langridge P, Brettschneider R, Lazzeri P, Lorz H (1992) Transformation of cereals via *Agrobacterium* and the pollen pathway: a critical assessment. Plant J 2:631–638
Lazo GR, Pascal AS, Ludwig RA (1991) A DNA transformation-competent *Arabidopsis* genomic library in *Agrobacterium*. Bio/tech 9:963–967
Luo Z-X, Wu R (1989) A simple method for the transformation of rice via the pollen-tube pathway. Plant Mol Biol Rep 7:69–77
Martin T, Wohner RV, Hummel S, Willmitzer L, Frommer WB (1992). The GUS reporter gene system as a tool to study plant gene expression. In: Gallagher SR (ed) GUS protocols: the GUS gene as a reporter of gene expression. Academic Press, San Diego, pp 23–43
Mentewab A. Letellier V, Marque C. Sarrafi A (1999) Use of anthocyanin biosynthesis stimulatory genes as markers for the genetic transformation of haploid embryos and isolated microspores in wheat. Cereal Res Commun 27:17–24
Meyer P (1995) Understanding and controlling transgene expression. Trends Biol Technol 13: 332–337

Meyer P, Niedenhof I, ten Lohuis M (1994) Evidence for cytosine methylation of non-symmetrical sequences in transgenic *Petunia hybrida*. EMBO J 13:2084–2088

Murashige T, Skoog F (1962) A revised medium for rapid growth and bioassays with tobacco tissue cultures. Physiol Plant 15:473–479

Ohta (1986) High-efficiency genetic transformation of maize by a mixture of pollen and exogenous DNA. Proc Natl Acad Sci USA 83:715–719

Ottaviani MP, Smits T, Hanisch ten Cate HH (1993) Differential methylation and expression of the β-glucuronidase and neomycin phosphotransferase genes in transgenic plants of potato cv. Bintje. Plant Sci 88:73–81

Prols F, Meyer P (1992) The methylation patterns of chromosomal integration regions influence gene activity of transferred DNA in *Petunia hybrida*. Plant J 2:465–475

Qing CM, Fan L, Lei Y, Bouchez D, Tourneur C, Yan L, Robaglia C (2000) Transformation of Pakchoi (*Brassica rapa* L. ssp. *chinensis*) by *Agrobacterium* infiltration. Mol Breed 6:67–72

Ramaiah SM, Skinner DZ (1997) Particle bombardment – a simple and efficient method of alfalfa (*Medicago sativa* L.) pollen transformation. Curr Sci 73:674–682

Sanford JC, Skubik KA (1986) Attempted pollen-mediated transformation using Ti plasmids. In: Mulcahy DL, Bergamini-Mulcahy H, Ottaviano E (eds) Biotechnology and ecology of pollen. Springer, Berlin Heidelberg New York, pp 1–82

Senior IJ, Dale PJ (1996) Plant transgene silencing – gremlin or gift? Chem Indus 19:604–608

Southern EM (1975) Detection of specific sequences among DNA fragments separated by gel electrophoresis. J Mol Biol 98:503–517

Tjokrokusumo D, Heinrich T, Wylie S, Potter R, McComb J (2000) Vacuum infiltration of *Petunia hybrida* pollen with *Agrobacterium tumefaciens* to achieve plant transformation. Plant Cell Rep 19:792–797

Trieu AT, Burleigh SH, Kardailsky IV, Maldonado-Mendoza IE, Versaw WK, Blaylock LA, Shin HS, Chiou TJ, Katagi H, Dewbre GR, Weigel D, Harrison MJ (2000) Transformation of *Medicago truncatula* via infiltration of seedlings or flowering plants with *Agrobacterium*. Plant J 22: 531–541

Zheng J-Z, Wang D-J, Wu T.-Q, Zhang J, Zhou W-J, Zhu X-P, Xu N-Z (1994) Transgenic wheat plants obtained with pollen-tube pathway method. Sci China 37:319–325

4 Onion, Leek and Garlic Transformation by Co-Cultivation with *Agrobacterium*

C.C. EADY

4.1 Introduction

Allium species are difficult to transform. Because of the difficulties, the genus is amongst the last commercially important vegetable genus for which gene transformation protocols are being developed. The protocols outlined in this chapter are still in their infancy and in the case of leek and garlic, they are the only successful reports of *Agrobacterium*-mediated transformation for those particular species (at the time of going to press). For onions, only two protocols have been published. Of these, only the one described below (Sects. 4.2.1 and 4.2.2) has been routinely used to produce transgenic onions containing several different traits. Therefore, the reader should be aware that while every effort has been made to accurately report the technology that is currently being used, there is every likelihood that these protocols can be improved upon or that new protocols may arise that will supersede what is reported here. This chapter deals with *Agrobacterium*-mediated transformation and does not cover direct gene transfer methods, such as biolistics and cell fusion-mediated transformation or interspecific gene integrations. For information on these gene transfer techniques in *Allium* species, the reader is referred to Eady (1995, 2002a,b), Buitveld (1998) and Kik (2002). This chapter does not cover the history of *Allium* transformation, gene delivery, gene regulation, and cell culture research, or the potential applications and risks of the technology to alliums. These have been covered in previous reviews (Eady 1995, 2002a,b). This chapter outlines the current state of the technology and how it is being applied. Rather inevitably, in this day and age, it is possible that some current applications have been omitted as some research may be subject to nondisclosure due to its commercial sensitivity. There may also be other cases for which published results are not yet available.

4.1.1 Current Applications of *Allium* Transformation Technology

4.1.1.1 Physiological Studies

The *Allium* genus has particularly interesting sulfur and carbohydrate pathways (reviewed by Block 1992; Hell 1997; Leukstek and Saito 1999; Randle and Lancaster 2002 for sulfur, and Darbyshire and Steer 1990 for carbohydrate).

Molecular Methods of Plant Analysis, Vol. 23
Genetic Transformation of Plants

These are responsible for some of the unique neutraceutical qualities ascribed to garlic and onions and this has created a great deal of interest in understanding these pathways. Gene discovery programs to identify the candidate genes playing a role in sulfur or carbohydrate pathway regulation of alliums have been established in several laboratories around the world (Van Heusden et al. 2000; Galmarini et al. 2001; McCallum et al. 2001). Candidate genes identified as involved in these pathways are being used to build RNA interference (RNAi) constructs in order to silence specific genes in the onion genome so that their role in sulfur and carbohydrate regulation can be determined. Plants are currently being developed with RNAi constructs directed against serine acetyl transferase (SAT1) and γ-glutamyl cysteine synthetase (GGCS) genes in the S pathway and sucrose phosphate synthetase (SPS) in the carbohydrate pathway (Eady and McCallum, unpubl.). To date, only one gene, the alliinase gene, has been silenced using antisense technology (Eady 2002a). The effect of silencing this gene is described in Section 4.2.5.3. In parallel to this research, Kondo et al. (2000) have recently developed a garlic transformation procedure, described in Section 4.4, with the aim of manipulating the alliinase levels (Suzuki and Kondo, pers. comm.).

4.1.1.2 Herbicide Resistance

Herbicide-resistant onion germplasm has now been developed from several commercially important lines. This work has concentrated on the use of the CP4-derived gene constructs to confer resistance to the systemic herbicide glyphosate (Eady et al. 2002a). Initial results indicate that the gene is stable and constitutively expressed and causes no obvious phenotypic alteration to the onion. Savings in herbicide usage of up to 75% for this crop have been projected (Eady 2001). In addition, glyphosate is a short-lived low toxicity herbicide compared with many of the persistent toxic herbicides that are currently used.

For experimental purposes, gene constructs containing the *bar* gene have also been introduced to demonstrate that plants resistant to the contact herbicide phosphinothricin can also be produced (Eady et al. 2002a). Stable inheritance of this gene into the first filial generation has also been demonstrated (Eady et al. 2002b).

4.1.1.3 Antimicrobial Resistance

Work has begun on the introduction of oxalate oxidase-containing gene constructs in an effort to produce transgenic onions that can neutralize the oxalic acid produced by invading fungal pathogens, such as *Sclerotium cerpivorum* Berks., the cause of onion white rot. Such work is hindered, at present, by the lack of rapid in vitro assays that can reliably demonstrate the effectiveness of transgene products against specific pathogens. Such systems are being developed for onion (Hunger et al. 2002).

The introduction of synthetic magainin gene constructs to express small antimicrobial peptides in onion is also being undertaken (unpubl.). These peptides disrupt the integrity of invading microbes, thus hindering their progress through the plant. Magainin-expressing potato plants, grown in the field, have been produced with a significant tolerance to soft rot (*Erwinia* spp.; Barrell 2001). It is expected that transgenic onion plants will show a similar response to onion soft rots.

As an alternative approach, our research team at Crop & Food Research, Lincoln, New Zealand, is attempting to silence root alliinase expression as there is strong evidence suggesting that alliinase activity is responsible for the production of volatile sulfur compounds that are released into the soil. These compounds stimulate germination of dormant fungal sclerotia (Bansal and Broadhurst 1992). If alliinase expression can be switched off in the root, and lines with significantly reduced alliinase have been produced (Eady et al. 2000a), then it should be possible to reduce the release of sulfur compounds into the soil and consequently inhibit sclerotial germination. Unfortunately, many soil microbes are capable of degrading the sulfur precursors (King and Coley-Smith 1969) that may accumulate in onion roots if alliinase is absent. In vivo experimentation is required to clarify whether or not this approach has potential.

4.1.1.4 Insect Resistance

Crop alliums can suffer damage from many insect pests. These pests will vary, depending on the part of the world where the crop is grown (Soni and Ellis 1990). However, some of these pests are generalist grazers, such as thrips, and it is likely that a simple genetic basis of resistance to such pests will be difficult to develop. For others, such as the beet armyworm (*Spodoptera exigua* Hubber) and onion maggot, it should be possible to develop effective transgenic approaches (Eady 2002a). Work has been initiated in this area at Plant Research International (PRI), Wangeningen (Zheng et al. 2000, 2001) and several shallot plants containing *Bt cry* gene constructs have been produced (Kik, pers. comm.). It remains to be seen whether this will deter beet armyworm feeding.

4.2 Onion Transformation Protocols

The transformation procedure outlined below is used routinely in the Crop & Food Research laboratory at present and is based on that developed by Eady et al. (2000b). It is inefficient, but reliable and should be amenable to improvement. The only other onion transformation technique currently available is based on the use of selected lines of mature zygotic cultures. For details about this technique, the reader is referred to Zheng et al. (2001).

4.2.1 Transformation Using Antibiotic and Visual Selection

4.2.1.1 Bacterial Strain and Plasmids

Agrobacterium tumefaciens strain LBA4404 with the binary vector *pBIN* or pCambia derivatives, and possessing within its T-DNA the selectable genes *nptII* or *hyg* and the visual reporter genes *uidA* or *m-gfpER*, has been used in *Allium* transformation experiments. In experiments to date, expression has been driven by either the *CaMV35s* promoter or *nos* (Eady et al. 2000b; Zheng et al. 2001).

4.2.1.2 Transformation Procedure (Modified from Eady et al. 2000b)

- Day 1: Initiate *Agrobacterium* cultures by inoculating 50 ml of LB media containing appropriate selective agents, in a 100-ml flask, with 1 ml of frozen stock (0.7 ml of actively growing culture plus 0.3 ml of 50% glycerol stored in 1-ml aliquots at –80°C). Grow overnight at 28°C, ~125 rpm.

  Use field-grown umbels of *Allium cepa* L. as a source of plant material. Isolate, using fingernails or tweezers, approximately 12 g of immature seed containing 2–5-mm embryos (large translucent embryos seem to be the most responsive). Seeds should have recently turned black, but should still have a liquid endosperm. Store the isolated seeds overnight at 4°C in a 50-ml vial and cover with wet tissue.
- Day 2: Replenish the *Agrobacterium* culture with an equal volume of LB containing antibiotic and 100 μM acetosyringone (virulence gene-inducing factor), and grow for a further 4–5 h. When ready, adjust the optical density to about 0.4 at 550 nm by the addition of LB. After the immature embryos have been isolated (see below), *Agrobacteria* are pelleted in 15-ml Falcon tubes by centrifugation at 4000 rpm and resuspended in an equal volume of liquid plant culture media, P5, containing 200 μM acetosyringone.

  Wash isolated immature seeds for 1 min in ethanol and then surface sterilize in 50 ml of 30% household bleach solution, plus 1 drop of Tween–100, for 30 min with occasional shaking. After washing four times in sterile water and draining for 2 min, isolate immature embryos from the seed by stereo microscope (Eady et al. 1998). Isolate batches of 50 embryos, cut into ~1-mm lengths and hold in 50 μl of liquid P5 (Eady et al. 1998) in an Eppendorf tube until ~1000 embryos have been isolated. Isolating healthy embryos and minimizing damage to the embryos is crucial in order to get a good response. Add 0.4 ml of *Agrobacteria* to each Eppendorf tube and vortex for 30 s. Pierce a hole in the lid of the Eppendorf tube with a needle and then place under vacuum (~25 mmHg) for 30 min. Siphon off the liquid using a 200-μl pipette and carefully transfer the immature embryo tissue pieces to filter paper. Leave to dry, in a clump, for approximately 1 min before transferring to P5 media and carefully plating out with fine tweezers (two batches per Petri dish). It is important to make sure that the P5 media is dry on the surface; wet media causes water to pool around the embryo and this severely reduces embryo survival. After transfer, label dishes sealed with household Gladwrap and place in the dark at 24–26°C in the growth room.

- Day 7: After 5-days co-cultivation, transfer embryo pieces to P5 containing timentin (200–250 mg/l) to kill the *Agrobacterium*, and geneticin at 12 mg/l to select for transgenic tissue. Culture embryo pieces in the dark under the same conditions as described above, transferring to fresh medium every fortnight. For selection on geneticin, we have found it best to select for 4 weeks at 12 mg/l, followed by 8 weeks at 20 mg/l. The timentin can normally be dropped out at week 6–8. However, if this is done, it is necessary to watch carefully for any latent *Agrobacteria* overgrowth. If this occurs, clean tissue must be immediately placed in media containing timentin.

  From day 7 onwards, it is possible to observe any transgenic material containing the *m-gfpER* gene using a fluorescence microscope by observation of the tissue under 440–480 nm excitation and 510 nm emission (Leffel et al. 1997; Eady et al. 2000b). If *uidA* visual reporter gene is used, then tissue has to be sacrificed and stained (Eady et al. 1996; Zheng et al. 2001).
- Week 3: After 2 weeks of selection, transfer actively growing material to regeneration medium (Eady et al. 1998) containing 20 mg/l geneticin. (This time period can be reduced to 8 weeks for some tissue pieces if the visual *m-gfpER* marker is also used to identify transgenic tissue.) Maintain shoot cultures for 12 weeks under 16-h day length by subculture every 3 weeks.
- Week 19 onwards: Transfer developing shoots to 1/2 MS media (Murashige and Skoog 1962) plus 20 mg/l geneticin and maintain as above to induce rooting. In vitro culture response is varied and some transgenic sectors proliferate to produce hundreds of shoots on regeneration media while others fail to produce shoots or only produce a few. Maintenance for long periods (>6 months) in vitro results in a decline in regeneration capability and an increase in the degree of hyperhydricity exhibited by the cultures (Eady et al. 1998). It is possible to induce in vitro bulbs by subculturing the plantlets on 1/2 MS plus 120 g/l of sucrose (Seabrook 1994).

4.2.2 Transformation Using Herbicide Selection

Transgenic onion plants containing genes that confer resistance to the herbicides glyphosate and phosphinothricin, respectively, have been introduced into onion. The process essentially follows that outlined in Section 4.2.1, with the modifications described below. The process is described in detail in Eady et al. (2002a).

4.2.2.1 Bacterial Strain and Plasmids

To generate glyphosate-tolerant plants, the following strain was used: *Agrobacterium tumefaciens* strain ABI, containing a binary vector which included the *CP4* gene within the T-DNA (kindly supplied by Monsanto Corp.). To generate plants tolerant to phosphinothricin, the following strain was used: *Agrobacterium tumefaciens* strain LBA4404 containing a binary vector pCambia 3301 (from CSIRO, Canberra) with the *bar* gene within the T-DNA.

4.2.2.2 Transformation Procedure

The following modifications were made to the procedure outlined in Section 4.2.1. For the selection of glyphosate tolerant tissue, co-cultivated embryos are initially transferred to P5 media containing 0.025 mM glyphosate and 250 mg/l timentin. The glyphosate level is raised to 0.05 mM after 6–8 weeks selection and maintained at this level throughout selection and regeneration. For the glyphosate-tolerant plants, it is necessary to maintain timentin levels through transfer to shoot media as the ABI strain is more virulent than LBA4404 and so is prone to overgrowth.

For the selection of phosphinothricin-tolerant tissue, co-cultivated embryos are initially transferred to P5 media containing 5 mg/l phosphinothricin and 250 mg/l timentin. The phosphinothricin is maintained at this level throughout selection and regeneration.

4.2.3 Ex-Flasking and Growth in Containment

Transferring the primary transformant from in vitro culture to the glasshouse is often a technically difficult process. Fortunately, onion plantlets in culture are quite robust and there are numerous reports of successful transfer to the glasshouse (Novak 1990; Eady 1995). Rooted plants are carefully washed to remove any attached agar and transferred to a 1:1:1 perlite:bark:compost mix containing 400 g Dolomite, 300 g Osmocote, 75 g Superphosphate and 50 g sulphate of potash per 50 l in the glasshouse (12 h, 12–23°C day, 12 h, 4–16°C night). Ex-flasking has proved relatively easy in onions. However, maintenance within transgenic containment facilities can be a problem. For correct chilling and flowering, it is preferable to grow them within a transgenic shade house. In wet winters and in spring the wet still air in such shade houses is perfect for scarab fly and a range of bacterial and fungal pathogens. In these circumstances, care must be taken in order to avoid losing important material. Bulbing in glasshouse-grown plants is induced naturally by increasing day length. After 50% of the tops have fallen, bulbs are lifted and air-dried. Bulbs are rested for 2 months in a warm dry environment and then either cold-stored to induce floral meristems prior to planting in the spring or replanted in large pots and grown in a containment shade house over winter, spring and summer in order to produce flowers.

Transgenic plants are self fertilised by bagging the umbel (within a micro-perforated plastic bread bag) as the flowers start to open. Greenbottle flies or ready-to-hatch pupae (about 20–50) are then introduced into the bag and left until all flowers have finished opening (occasionally the introduction of additional flies is necessary). Crossing onto nontransgenic plants is achieved by flowering plants in 1 × 1 × 2-m-high cages. Flies or bees were introduced to effect pollination.

After all the flowers are pollinated the umbel is left to ripen. As the umbels ripen, the ovary walls containing the seed start to crack open. At this point, the umbels are harvested and air-dried in large 1-mm mesh bags for 1 month. Seeds can then be separated from the chaff and either stored (viability drops dramatically in onions unless stored properly) or germinated immediately for further investigation.

4.2.4 Transgene Detection

Initially, the presence of the transgene in putative transgenic onion tissue was screened using PCR in order to amplify specific fragments of a particular transgene. For the *bar* gene we have been unable to specifically detect the presence of the *bar* transgene fragment without also obtaining very faint positive bands within nontransgenic onion samples. Despite a large difference in the degree of amplification (at least tenfold), this uncertainty has led us to conclude that in onion, under our laboratory conditions, PCR cannot be used to conclusively demonstrate the presence of transgene fragments. However, it is routinely used to screen for glyphosate *nptII* and *mgfpER* genes. For specific PCR conditions for these genes, refer to Eady et al. (2000b, 2002a,b).

Currently, transformants are usually confirmed by Southern analysis following the protocol of Eady et al. (2000b). However, Southern analysis is expensive and time-consuming. In cases where clonal propagation of a particular transgenic culture is pertinent, it is necessary to confirm that all clones are derived from the same transformation event. In such a case where over 100 clones could easily be produced, Southern analysis would be inappropriate. For this scenario, and for the rapid easy fingerprinting of transgenic lines, our group is developing the application of thermal asymmetric interlaced PCR (TAIL-PCR; Liu and Whittier 1995) strategies to enable the detection of transgenic lines. TAIL-PCR consists of three sequential PCR reactions using nested primers from the known inserted DNA sequence directing amplification towards the unknown flanking region, and a particular arbitrary degenerate primer to amplify back from the unknown region. We have designed TAIL-PCR-specific nested primers and identified an arbitrary degenerate primer that can efficiently amplify most integrations of a pBin19-based T-DNA in an *Allium* genome. DNA from individuals of a population of 66 plants was used as template for a TAIL-PCR, using pBIN19 night border primers Apo4RB1, 2, 3, and arbitrary degenerate primer AD2 (Catanach, unpubl.). From the 67 plants, 47 (70.1%) gave results distinctive of a TAIL-PCR step-down pattern. Sixteen different classes of product signatures were found and, in many cases, multiple plants from the same transformation plate were found to be of clonal origin. In addition, plants from the same transformation plate with different TAIL-PCR signatures were found indicating that different transformation events could also arise from the same transformation plate.

A further benefit of TAIL-PCR over Southern analysis is that it enables sequencing of genomic DNA that flanks T-DNA inserts. Sequences of TAIL-PCR products have thus far demonstrated very precise integration at both the left and the right borders of the T-DNA in *Allium*. Furthermore, these sequences may be useful in identifying PCR probes for two purposes:

1. Transgenic hybrid onion seed is being developed by crossing a nontransgenic open pollinated parental line with a particular transgenic parental plant carrying a single transgene in the hemizygous state. Some resulting seed on the nontransgenic parents will be hemizygous for the transgene and can be selected to give F_1 heterozygous individuals containing the transgene. Selfing these individuals will produce homozygous, hemizygous and null F_2 progeny with respect to the transgene. In order to rapidly identify the homozygous individuals from the hemizygous individuals, so that a heterozygous population, homozygous for the transgene, can be identified for bulk seed development, it is necessary to have a rapid seedling test for homozygosity vs. hemizygosity. Onions take 18 months to reach sexual maturity; therefore segregation analysis causes a significant delay during the route-to-market. The identification of homozygous individuals at the seedling stage using PCR primers or proves within the flanking sequence will prevent this delay.
2. PCR primers to detect sequences that flank T-DNA will provide a means of cultivar identification and presumably ownership.

4.2.5 Transgene Expression and Stability

4.2.5.1 Visual Reporter Genes

m-gfpER and *uidA* genes under the control of the *CaMV35s* promoter have been introduced into onion. Transformants regenerated under selection indicated that both genes are switched on in tissues required during culture and regeneration (Eady et al. 2000b; Kondo et al. 2000; Zheng et al. 2001). The profile of

CaMV35s expression in primary transformants and first filial generations has been determined extensively using the *m-gfpER* gene (Eady et al. 2000b, 2002b; Eady 2002a, b). The conclusions from these studies indicate that transgenes in onion, despite its large genome, are expressed in the same general way as transgenes in other species.

Twelve independent transformants from initial studies have been selfed and F_1 plants produced. Initial results indicate that the transgene is usually inherited in a normal Mendelian fashion and that offspring are phenotypically normal (Eady 2002a,b; Eady et al. 2002b).

4.2.5.2 Expression of Herbicide Resistance

Onion plants containing a *CaMV35s-bar* gene construct and the constitutively expressed glyphosate resistance gene, CP4, have been produced. All plants confirmed by Southern analysis as containing one or two copies of the transgene have exhibited a strong tolerance to the respective herbicides Buster and Roundup.

Buster is a contact herbicide and initially leaves were painted with a 0.5% solution of Buster. Plants demonstrating resistance to Buster were then sprayed with commercially recommended concentrations of the herbicide for general purpose weed control to confirm resistance. The level of resistance achieved indicates that the commercial production of transgenic onions containing a *bar* resistance gene is a feasible option for weed control in onions.

Similarly, glyphosate-tolerant plants produced to date have proven to be tolerant to twice the recommended field application rates required for general weed control. F_1 seed has recently been produced from these plants and they are currently being tested for commercial development. It has been estimated that the deployment of glyphosate-tolerant lines in New Zealand could substitute the application of up to 15 l of mainly toxic and persistent herbicides per season for approximately 4.5 l of low toxicity short-lived glyphosate which equates to an economic saving of about US$ 250/ha (Eady 2001).

4.2.5.3 Antisense Alliinase Gene Expression

Three sets of transgenic onion plants containing antisense alliinase gene constructs have recently been produced. The presence of the construct in some transgenic plants has been determined by Southern blot detection of flanking T-DNA sequences (Eady et al. 2002c). However, the precise number of independent transgenic lines has not yet been determined as TAIL-PCR analysis (see Sect. 4.2.4) has indicated that some material which was initially thought to be clonal is actually of independent transgenic origin or escape nontransgenic regenerants. The three sets of plants contain a *CaMV35s*-driven antisense root alliinase gene (Lancaster et al. 2000); a *CaMV35s*-driven antisense bulb alliinase, initially isolated by Clark (1993); and a bulb alliinase promoter-driven antisense bulb alliinase (Gilpin et al. 1995). All constructs were developed by Pither-Joyce (unpubl.). Primary transgenic root and bulb material has been analysed for alliinase levels (using Western analysis) and alliinase activity (by indirect analysis of pyruvate production, Randle and Bussard 1993). To date, initial results are somewhat confusing due to several compounding factors. These include the precise nature of the transgenic line, variation in the size and physiological status of the primary transformants, sensitivity of the assays, and the fact that often the antisense approach is inefficient with only about one in ten lines producing a good silencing response (Smith et al. 2000). It appears that the antisense root alliinase constructs have little or no effect on reducing root alliinase levels. Several repeated assays have now indicated that none of the antisense root lines have significantly reduced root alliinase levels compared to control plants. If the antisense technique is working, then this would suggest that the root alliinase sequence described by Lancaster et al. (2000), with only ~50% homology with the other alliinase sequences (Randle and Lancaster 2002), is not in fact the source of a rate-limiting root alliinase. Results from the antisense bulb alliinase lines have been much more encouraging and three lines were produced with barely detectable bulb alliinase levels and activity. F_1 plants from these lines have been produced and further analysis of alliinase expression and the physiological consequences of reduced transcript level are being examined. However, this has been compounded by the poor survival of transgenic plants in a particularly unfavourable season. Of the three lines, one has been lost, one remains as a single positive F_1 seedling, and fortunately the third line has produced several F_1 seedlings.

4.3 Leek Transformation

The onion transformation protocol was successfully applied to leek immature embryos to demonstrate for the first time stable transformation of leek. The following modifications on and comments to the onion protocol have been made (Davis et al., submitted).

Immature umbels of leak were more difficult to extract as the ovary wall tissue was tougher and more slippery than onion ovary wall tissue. In addition, when the seeds were at the correct developmental stage, they bleached upon sterilization. However, this did not affect the viability of the embryo. Initial embryo treatment and co-cultivation exactly followed the onion transformation protocol. Selection and regeneration was performed using only one level of geneticin selection (12 mg/l). It should be noted that leek embryo response to P5 media was markedly different to that of onion. The majority of leek embryos rapidly elongated and failed to form an embryogenic culture. This reduction in embryogenic response was most likely responsible for the reduction in transformation efficiency observed in leek as compared to onion.

Initial gene transfer, as determined by observations of transient GFP expression, appeared similar to that observed in commercial lines of onion, but less than that observed in the open pollinated Canterbury Longkeeper lines with which the onion transformation procedure was developed.

From about 5000 initial embryos, 2 transgenic leek lines have been produced. Multiple shoot cultures from these lines have been used to develop ex-flasking procedures for leek. Leek plants are not as robust as onion, probably due to their V-shaped as compared to circular leaf structure. They also appear to require more specific day length requirements for continued growth after ex-flasking (entering autumn with a shortening 13-h day). As a consequence of this, initially most ex-flasked plants died. However, currently about five transgenic plants have been successfully ex-flasked to the glasshouse and are being grown to maturity.

4.4 Garlic Transformation Protocol

The transformation procedure outlined below is based, with kind permission, on the method of Kondo et al. (2000). It is crucial to select highly regenerative callus lines for successful garlic transformation using this procedure. In addition to the procedure outlined below, our group has undertaken a preliminary investigation into the feasibility of transforming immature embryos of true seed garlic. To date, we have managed to recover stable transgenic root cultures. We are confident that, with healthy immature embryos and modified culture media, garlic, like leek and onion, can be transformed using this technique. Garlic has also been transformed by biolistic-mediated transformation (Alba Estela Jofre-Garfias, pers. comm.). However, the exact nature and efficiency of this transformation system has yet to be determined.

4.4.1 Bacterial Strain and Plasmids

Agrobacterium tumefaciens strain EHA101 (Hood et al. 1986) with the binary vector *pIG121Hm*, possessing within their T-DNA the selectable genes *nptII* and *hpt* and the visual reporter gene *uidA*, have been used in garlic transformation experiments.

4.4.2 Transformation Procedure

The transformation protocol relies on the previously developed procedure for highly efficient regeneration from garlic protoplasts (Hasegawa et al. 2002). Highly regenerative callus lines obtained from protoplasts were used as plant material for garlic transformation (Kondo et al. 2000).

- Day 1: Initiate *Agrobacterium* cultures by inoculating 200 ml of LB media containing appropriate selective agents, in a 500-ml flask, with 100 µl of frozen stock. Grow overnight at 28°C, ~154 rpm until turbidity reaches 0.6 OD_{600}.
- Day 2: *Agrobacteria* are pelleted by centrifugation at 8000 rpm for 10 min and resuspended in 100 ml of liquid plant culture media, MS, containing 1% glucose and 100 µM acetosyringone. Approximately 250 callus clumps (~1 cm in diameter) are submerged in the suspension of *Agrobacteria* for at least 5 min. Calluses are strained from bacterial suspension with a tea strainer and transferred to a paper towel to blot away the excess bacterial suspension. Calluses are carefully transferred (do not bury) to the MS medium (pH 5.2) solidified with 0.5% Gelrite containing 1% glucose, 1 mg/l 2,4-D and 100 µM acetosyringone. They are then co-cultivated in darkness for 3 days at 22°C (the culture periods and temperature of co-cultivation are important factors).
- Day 5: Calluses are washed five times with sterile water containing 150 mg/l timentin, blotted to remove excess liquid, and cultured for 2 months on CPM (callus proliferation medium solidified with 0.5% Gelrite) consisting of MS medium containing 1 mg/l 2,4-D, 150 mg/l timentin and 10–40 mg/l hygromycin. The concentration of hygromycin is initially at 10 mg/l, but then increased every 2 weeks to 20, 30 and 40 mg/l. Kanamycin is not suitable for the selection of transgenic garlic cells.
- Week 9: Calluses are transferred to SIM (shoot-inducing medium solidified with 0.5% Gelrite) consisting of MS medium containing 2 mg/l BA, 0.02 mg/l NAA, 150 mg/l timentin, and 20 mg/l hygromycin. During selection many calluses become brown or grey. However, it is very difficult to distinguish transgenic tissue accurately from proliferated calluses without staining. Therefore, all surviving calluses are maintained for about 5 months under 12-h day length by subculture every 2 weeks.
- Week 29 onwards: Developing shoots are transferred to MS medium containing 150 mg/l timentin and 20 mg/l hygromycin without plant growth regulators for rooting. It is possible to induce bulbs by subculturing the plantlets on MS containing 12% sucrose under constant fluorescent illumination.

  Plantlets were acclimatized in the glasshouse and then transferred to pots containing soil and vermiculite. The transgenic nature of the plants was determined by Southern analysis.

4.5 Concluding Remarks

Methods for the transformation of important *Allium* species are still very much in their infancy and transformation efficiencies of less than 1% of starting material is still the norm. The results obtained so far suggest that the difficulties faced in obtaining transgenic alliums lie with the effective production and selection of regeneration-competent cells that are capable of taking up DNA. DNA uptake via *Agrobacterium* or bombardment can be as efficient as that observed in many other plant transformation systems. DNA integration, when it occurs and is successfully maintained, also appears no different to that obtained in other plant transformation systems. Despite the difficulties of *Allium* transformation, their importance as a flavorsome crop (Randle and Lancaster 2002) which confers nutritional and health benefits (Keusgen 2002), has inspired some laboratory-hardened scientists to persevere. These scientists are developing systems and transgenic plants that will pave the way for biochemists to manipulate the key enzymes involved in sulfur and carbohydrate metabolism. This will further our understanding of the role of these pathways and enable us to identify which are important for the plant's normal physiology, and which are crucial components contributing to the neutraceutical value of these crops.

In addition, the ability to transform elite *Allium* cultivars will enable the introgression of traits from outside the *Allium* gene pool that can confer useful agronomic traits. Pest and disease resistance traits, such as those being applied to many other crops, have the potential to reduce the levels of pesticide and fossil fuels required for this crop and pave the way for the sustainable and efficient production of *Allium* crops.

Acknowledgements. I should like to thank Tracy Williams and Katherine Trought for their critical editing of this manuscript, Hajime Hasegawa and Masahiko Suzuki for the garlic transformation protocol, Chris Kik and Si-Jun Zheng for correspondence about shallot transformation, Sheree Davis and Fernand Kenel for the leek transformation work, Andrew Catanach for his work on TAIL PCR, and John McCallum, Meeghan Pither-Joyce and Martin Shaw for their molecular and biochemical skills.

Lastly, thanks also go to Crop & Food Research, the New Zealand Foundation for Research, Science and Technology (CO2×0017) and Seminis Vegetable Seeds, Woodland, CA for making so much of this work possible.

References

Bansal RK, Broadhurst PG (1992) An evaluation of *Allium* germplasm for resistance to white rot caused by *Sclerotium cerpivorum* Berk. NZ J Crop Hortic Sci 20:362–365

Barrell P (2001) Expression of synthetic magainin genes in potato. PhD Thesis, Lincoln University Christchurch, New Zealand

Block E (1992) The organosulfur chemistry of the genus *Allium* – implications for the organic chemistry of sulfur. Agnew Chem Int Ed Engl 31:1135–1178

Buiteveld J (1998) Regeneration and interspecific somatic hybridization in *Allium* for transfer of cytoplasmic male sterility to leek. Wageningen Agricultural University 127. Wageningen Agricultural University, Wageningen

Clark SA (1993) Molecular cloning of a cDNA encoding alliinase from onion (*Allium cepa* L.). PhD dissertation, University of Canterbury, Christchurch, New Zealand

Darbyshire B, Steer BT (1990) Carbohydrate biochemistry. In: Rabinowitch HD, Brewster JL (ed) Onions and allied crops, vol II. Agronomy, biotic interactions, pathology and crop protection. CRC Press, Boca Raton, pp 1–17

Eady CC (1995) Towards the transformation of onions (*Allium cepa*) (review). NZ J Crop Hortic Sci 23:239–250

Eady CC (2001) How green are your onions? Grower 56(11):23–24

Eady CC (2002a) Genetic transformation of onions. In: Rabinowitch HD, Currah L (ed) Allium crop science: recent advances. CABI Publishing/CAB International, Wallingford, UK, pp 199–144

Eady CC (2002b) The transformation of onions and related alliums. In: Khachatourians GG, McHughen A, Scorza R, NipW-K, Hui YH (eds) Transgenic plants and crops. Dekker, New York, pp 655–671

Eady CC, Lister CE (1998) A comparison of four selective agents for use with *Allium cepa* L. immature embryos and immature embryo-derived cultures. Plant Cell Rep 18:117–121

Eady CC, Lister CE, Suo YY, Schaper D (1996) Transient expression of uida constructs in in vitro onion (*Allium cepa* L.) cultures following particle bombardment and *Agrobacterium*-mediated DNA delivery. Plant Cell Rep 15:958–962

Eady CC, Butler RC, Suo Y (1998) Somatic embryogenesis and plant regeneration from immature embryo cultures of onion (*Allium cepa* L.). Plant Cell Rep 18:111–116

Eady CC, Pither-Joyce M, Farrant J, Shaw M, Reader J (2000a) In vivo suppression of alliinase in onion (*Allium cepa* L.). Sixth international congress of plant molecular biology, Quebec, Canada, 18–24 June 2000, abstract S03–33. International society for plant molecular biology, Université Laval

Eady CC, Weld RJ, Lister CE (2000b) *Agrobacterium tumefaciens*-mediated transformation and regeneration of onion (*Allium cepa* L.). Plant Cell Reports 19 (in press)

Eady CC, Davis S, Farrant J, Reader J, Kenel F (2002a) *Agrobacterium tumefaciens*-mediated transformation and regeneration of herbicide resistant onion (*Allium cepa* L.) plants. Ann Appl Bot (in press)

Eady CC, Reader J, Davis S, Dale T (2002b) Inheritance and expression of introduced DNA in transgenic onion plants (*Allium cepa* L.). Ann Appl Bot (in press)

Eady CC, Davis SD, Catanach A, Kenel F (2002c) Transgenic allium fingerprinting for quality control and breeding. 10th International Association for Plant Tissue Culture and Biotechnology, 23–28 June, Orlando, Florida (Abstract)

Galmarini CR, Goldman IL, Havey MJ (2001) Genetic analyses of correlated solids, flavor, and health-enhancing traits in onion (*Allium cepa* L.). Mol Genet Genom 265(3):543–551

Gilpin BJ, Leung DW, Lancaster JE (1995) Nucleotide sequence of a nuclear clone of alliinase (accession no l48614) from onion (PGR 95–125). Plant Physiol 110:336

Hasegawa H, Sato M, Suzuki M (2002) Efficient plant regeneration from protoplasts isolated from long-term, shoot primordia-derived calluses of garlic (*Allium sativum*). J Plant Physiol (in press)

Hell R (1997) Molecular physiology of plant sulfur metabolism. Planta 202(2):138–148

Hood EE, Helmer GL, Fraley RT, Chilton MD (1986) The hypervirulence of *Agrobacterium tumefaciens* A281 is encoded in a region of pTiBo542 outside of T-DNA. J Bacteriol 168:1291–1301

Hunger SA, McLean KL, Eady CC, Stewart A (2002) Seedling infection assay for resistance to *Sclerotium cepivorum* in onions (*Allium cepa*) and other allium species. Proceedings of the 55th New Zealand Plant Protection Society Conference. www.hortnet.co.nz/publications/NZPPS

Keusgen M (2002) Health and alliums. In: Rabinowitch HD, Currah L (eds) Allium crop science: recent advances. CABI Publishing/CAB International, Wallingford, UK, pp 357–378

Kik C (2002) Exploitation of wild relatives for the breeding of cultivated *Allium* species. In: Rabinowitch HD, Currah L (eds) Allium crop science: recent advances. CABI Publishing/CAB International, Wallingford, UK, pp 81–100

King J, Coley-Smith J (1969) Production of volatile alkyl sulphides by microbial degradation of synthetic alliin and alliin-like compounds, in relation to germination of sclerotia of *Sclerotium cepivorum* Berk. Annu Appl Biol 64:303–314

Kondo T, Hasegawa H, Suzuki M (2000) Transformation and regeneration of garlic (*Allium sativum* L.) by *Agrobacterium*-mediated gene transfer. Plant Cell Rep 19:989–993

Lancaster J, Shaw M, Pither-Joyce M, McCallum J, McManus M (2000) A novel alliinase from onion (*Allium cepa* L.) roots: biochemical characterization and cDNA cloning. Plant Physiol 122: 1269–1279

Leffel SM, Mabon SA, Stewart CN (1997) Applications of the green fluorescent protein in plants. Biotechniques 23:912

Leustek T, Saito K (1999) Sulfate transport and assimilation in plants. Plant Physiol 120(3): 637–643

Liu Y-G, Whittier RF (1995) Thermal asymmetric interlaced PCR: automatable amplification and sequencing of insert end fragments from p1 and YAC clones for chromosome walking. Genomics 25:674–681

McCallum JA, Leite D, Pither-Joyce M, Havey MJ (2001) Expressed sequence markers for genetic analysis of bulb onion (*Allium cepa*. L.). Theor Appl Genet 103:979–991

Murashige T, Skoog F (1962) A revised medium for rapid growth and bioassays with tobacco tissue cultures. Physiol Plant 15:473–497

Novak FF (1990) Botany, physiology, and genetics. In: Rabinowitch HD, Brewster JL (eds) Onions and allied crops, vol I. Botany, physiology and genetics. CRC Press, Boca Raton, pp 233–250

Randle WM, Bussard ML (1993) Streamlining onion pungency analysis. Hortscience 28:60

Randle WM, Lancaster JE (2002) Sulphur compounds in alliums in relation to flavour quality. In: Rabinowitch HD, Currah L (eds) *Allium* crop science: recent advances. CABI Publishing/CAB International, Wallingford, UK, pp 329–356

Seabrook J (1994) In vitro propagation and bulb formation in garlic. Can J Plant Sci 74:155–158

Smith NA, Singh SP, Wang MB, Stoutjesdijk PA, Green AG, Waterhouse PM (2000) Gene expression – total silencing by intron-spliced hairpin RNAs. Nature 407(6802):319–320

Soni S, Ellis P (1990) Insect pests. In: Rabinowitch HD, Brewster JL (eds) Onions and allied crops, vol II. Agronomy, biotic interactions, pathology and crop protection. CRC Press, Boca Raton, pp 213–272

van Heusden AW, Shigyo M, Tashiro Y, Vrielink-van Ginkel R, Kik C (2000) AFLP linkage group assignment to the chromosomes of *Allium cepa* L-via monosomic addition lines. Theor Appl Genet 100(3–4):480–486

Zheng SJ, Henken B, Wietsma W, Sofiari E, Jacobsen E, Krens FA, Kik C (2000) Development of bio-assays and screening for resistance to beet armyworm (*Spodoptera Exigua Hubner*) in *Allium cepa* L. and its wild relatives. Euphytica 114(1):77–85

Zheng SJ, Khrustaleva L, Henken B, Sofiari E, Jacobsen E, Kik C, Krens FA (2001) *Agrobacterium tumefaciens*-mediated transformation of *Allium cepa* L.: the production of transgenic onions and shallots. Mol Breed 7(2):101–115

5 Electroporation Transformation of Barley

F. Gürel and N. Gözükirmizi

5.1 Introduction

Barley (*Hordeum vulgare* L.) is a member of the tribe Triticeae of the family *Poaceae* (von Bothmer et al. 1991), which was widely adapted to temperate regions of the world. Barley has both two- and six-rowed types according to spike morphology; intermediate types also exist. Due to its good adaptation to unfavorable climates, barley is produced in a significant quantity and consumed as livestock feed, food for man and malt brewery materials. In 2001, barley was harvested from 53,827,895 ha (FAO, http://apps.fao.org/page/collections) and yielded 25,723 hg/ha of grain in the world. Today, barley is not only the fourth most important cereal crop (after maize, wheat and rice) in the world, but also an excellent model plant for biochemists, physiologists, geneticists and molecular biologists (Shewry 1992) because of its inbred diploid nature, low chromosome number $2n = 2x = 14$ (von Bothmer 1992) and relatively small genome size. The barley genome contains nearly 5.5 pg of DNA per haploid nucleus, equivalent to approximately 5.3×10^9 bp (Bennet and Smith 1976). Ease of growth under laboratory conditions facilitates the development of molecular markers for construction of genetic maps (Becker and Heun 1995; Hayes et al. 1996; Manninen 2000; Ramsay et al. 2000; Williams et al. 2001), genome analysis (Michalek et al. 2002) and functional genomics studies (Öztürk et al. 2002; Sreenivasulu et al. 2002). Special attention has also been given to the establishment of tissue culture and development of gene transfer systems in order to obtain desirable genotypes (Lemaux et al. 1999).

During conventional breeding, heritable variations are created mainly by controlled crosses between adapted high yielding cultivars and breeding lines, specific traits may be introgressed from wild barley and landraces in back crossing programs (Nevo 1992). For decades, the main objective of barley conventional breeding programs has been to increase the yield while maintaining grain quality for feed, food and malting. For example, total protein, soluble protein extract, fine/coarse difference, diastatic power and alpha-amylase levels are the factors for evaluating malting quality in the brewery. For animal feed, indigestible material in the outer layers of the seed and fiber values affect grain quality. Improvement efforts in barley are also concentrated on producing resistant varieties to pathogens which include nearly 30 fungal (e.g., spot blotch, anthracnose, powdery mildew, eyespot, downy mildew, snowrot), bacterial (e.g., *Psuedomonas*, *Xhanthomonas*), viral (e.g., stripe mosaic, barley

Molecular Methods of Plant Analysis, Vol. 23
Genetic Transformation of Plants

yellow and barley yellow dwarf viruses), other organisms (e.g., fruit fly, wireworm, bibionids, heterodera, insect pests) and abiotic stresses (e.g., drought, salt, cold, heat; Dunwell 1986).

Gene transfer technologies can offer a suitable alternative for improving barley with desirable characteristics. Establishment of stable and regenerative tissue culture systems is a prerequisite for barley transformation. So far, many tissue culture protocols from mainly immature embryos (Breiman 1985; Thomas and Scott 1985; Goldstein and Kronstrad 1986; Jorgensen et al. 1986; Karp et al. 1987; Lührs and Lörz 1987; Rotem-Abarbanell and Breiman 1989) mature embryos (Lupotto 1984; Katoh et al. 1986; Ukai and Nishimura 1987; Ahloowalia 1987; Rengel 1987; Rotem-Abarbanell and Breiman 1989; Gözükırmızı et al. 1990), apical meristems (Cheng and Smith 1975; Weigel and Hughes 1985), anthers (Kao and Horn 1982; Huang and Sunderland 1982; Piccirilli and Arcioni 1991; Hoekstra et al. 1992), microspores (Köhler and Wenzel 1985; Kao et al. 1991), cell suspensions (Kott and Kasha 1984; Seguin-Swartz et al. 1984; Lührs and Lörz 1988; Müller et al. 1989; Lazzeri and Lörz 1990; Lührs and Nielsen 1992) and protoplasts (Lazzeri and Lörz 1990; Yan et al. 1990; Jahne et al. 1991a) have been developed. Among the tissues used for this purpose, the use of mature embryos has captured special attention because of its advantages to protoplast and cell suspension cultures which may require longer culture periods. In addition, mature embryos are easily and reliably germinated and regenerated, representing an ideal system for the production of barley plants. Other features are availability in every season, easy-handling and relative sterility of the tissue. Regeneration procedures are based mostly on callogenesis of mature embryos, however, prolonged callus phase leads to an increase in genetic abnormalities and loss of morphogenesis. Plant regeneration is possible via both organogenesis and somatic embryogenesis from cultured mature embryos of barley (Gözükırmızı et al. 1990).

Plant transformation is a significant development that may allow the release of fertile transgenic barley plants. *Agrobacterium*-mediated transformation which is routinely used for dicotyledonous plants has been problematic for cereal plants including barley and has succeeded only recently (Cheng et al. 1997; Tingay et al. 1997; Wu et al. 1998; Patel et al. 2000). Therefore, direct gene transfer techniques have been mostly used in barley transformation. Very early techniques tested in barley transformation were the electrophoresis of germinating seeds with DNA (Ahokas 1989) or incubation of embryos in a DNA solution (Töpfer et al. 1989). Following procedures relied on protoplast transformation by using either PEG-mediated (Junker et al. 1987; Lee et al. 1989; Lazzeri et al. 1991; Funatsuki et al. 1995; Kihara et al. 1998) or electroporation transformation (Teeri et al. 1989; Salmenkallio-Marttila et al. 1995). Beside protoplasts, microspore suspensions (Joersbo et al. 1990), isolated scutella (Hansch et al. 1996) and mature embryos (Gürel and Gözükırmızı 2000) have been demonstrated as suitable explants for electroporation. On the other hand, transient (Mendel et al. 1989; Kartha et al. 1989; Knudsen and Müller 1991; Chibbar et al. 1993; Harwood et al. 1995; Stiff et al. 1995; Hagio et al. 1995; Jensen et al. 1996; Koprek et al. 1996; Yao et al. 1997) and stable (Wan and Lemaux 1994; Jahne et al. 1994; Ritala et al. 1995) expression of transgenes were obtained in particle bombarded explants of barley. Macroinjections of DNA into floral tillers of barley also resulted in stable transformations (Mendel et al. 1990; Rogers and Rogers 1992). Another transformation method tested was microinjection of DNA into microspores (Olsen 1991; Bolik and Koop 1991) and zygotic protoplasts (Holm et al. 2000) which resulted in two plant lines expressing the marker genes.

There are still problems for high efficiency transformation of desired genotypes in barley. In this chapter, we present electroporation-mediated transformation of this plant with the detailed experimental steps of a procedure developed in our laboratory using intact mature embryos.

5.2 Background of Electroporation Procedures

Electroporation is a method of choice for direct gene transfer, because it is (1) rapid, (2) reliable, (3) cost-effective, and (4) nontoxic for recipient cells. This method was first developed for transformation of mammalian cells (Neumann et al. 1982) and then used in broader applications of medical and plant biotechnology for transferring DNA, RNA, protein or therapeutic agents into cells. The method is based on the opening of hydrophilic pores on the membranes of cells subjected to electrical pulses. Through the pores formed on bilayer membranes, any of the macromolecules may directly be transferred into protoplasts or intact cells. Permeability of cell membranes lasts for a few minutes at room temperature and about 1 h at 0°C.

Electroporation procedures were first developed for plants by using protoplast systems of dicotyledons and monocotyledons (Fromm et al. 1985; Shillito et al. 1985). Furthermore, Dekeyser et al. (1990) electroporated leaf bases of rice seedlings and confirmed the applicability of this method for other cereals and explant types. To date, mature (Xu and Li 1994) and immature embryos (Rao 1995), shoot apex explants (Padua et al. 2001) of rice; embryogenic callus (Zaghmout 1993) anther culture-derived embryoids (Gustafson et al. 1995) and immature embryos (Klöti et al. 1993; Sorokin et al. 2000) of wheat; type I callus, immature embryos (D'Halluin et al. 1992; Songstad et al. 1993) and suspension cell cultures (Laursen et al. 1994) of maize have been electroporated.

5.2.1 Pre- and Post-Electroporation Period

Tissue culture steps before and after electroporation have great importance for recovery of regenerated transformants. In a number of experiments, special attention was given to timing of electroporation in order to find the most appropriate stage of cell division. The time chosen is also crucial to electroporate cells and intact tissues with minimal damage. Several explant types such as tritordeum inflorescences (He et al. 2001) and scutella of barley embryos (Hansch et al. 1996) were electroporated after a 1-day pre-culture period. However, mature embryos which are known to have a lower proliferation needed at least a 3-day pre-culture period (Gürel and Gözükırmızı 2000). Enzymatic treatment of cultured tissues before electroporation has been widely used in many reports as a part of transformation procedure. This treatment seems to be necessary to obtain reasonable transient expressions of reporter genes, however, the amount of enzymes and incubation time should be adjusted as these may negatively affect tissue regeneration. For example, enzymatic or mechanical treatment of maize tissues and subsequent electroporation reduced the type I callus formation from immature embryos (D'Halluin et al. 1992). Commonly used commercial enzymes for this treatment are cellulase, driselase, macerozyme and pectolyase. Specifically, a combination of pectolyase Y-23 (0.5%), macerozyme R-10 (1.0%) and driselase (0.5%) promoted DNA transfer and did not affect embryogenic capacity in tissue electroporation of barley (Hansch et al. 1996).

The composition of electroporation buffer (EB) is formulated based on conductivity. High-conductivity buffer is mostly preferred in cereal electroporation. For example, electroporation buffer developed for both dicotyledonous and monocotyledonous protoplasts is a phosphate-buffered saline (Fromm et al. 1985) which contains (1×) 1.8 g/l NaCl, 0.2 g/l KCl, 0.2 g/l KH_2PO_4, 1.15 g/l Na_2HPO_4 and the concentration of 0.6× gave a higher transient expression of chloramphenicol acetyl transferase (CAT) reporter gene. In the same study, electroporation efficiency was tested in different Ca^{2+} ion concentrations (0, 2, 4, 6, 8, 10 mM $CaCl_2$) in a Hepes-buffered saline solution (10 mM Hepes, pH 7.2, 150 mM NaCl, 0.2 M mannitol) and 4 mM of $CaCl_2$ produced the maximal transient gene expression. In following studies, composition of the electroporation buffers included either high chloride concentrations (e.g., KCl, NaCl, $CaCl_2$, $MgCl_2$) or organic acids (e.g., aspartic acid monopotassium salt, glutamic acid monopotassium salt; Tada et al. 1990). Particularly, Tada et al. (1990) obtained a higher frequency of both transformation and colony formation in rice protoplasts by developing an organic acids-based buffer referred to as ASP (70 mM aspartic acid monopotassium salt, 5 mM calcium gluconate, 5 mM MES and 0.4 M mannitol, pH 5.8). The basic idea behind this was that the elimination of toxic Cl_2 gas released by the chlorides included in the KCl-based buffer and ASP resulted in a high number of transformants, increasing the cell survival. A parallel result was obtained in barley protoplasts in which transient gene expression of *uidA* (*E. coli* β-glucuronidase) gene was substantially increased when organic acid-based buffers were used instead of a chloride buffer (Salmenkallio-Marttila 1994). Consequently, the number of calli which grew on selection media was considerably increased with the use of organic acids. However, neither media affected the embryogenic capacity of electroporated intact tissues of barley, moreover, the use of KCl-based buffer was essential to obtain detectable gene expressions (Hansch et al. 1996). Osmoticum, which refers to compounds (e.g., fructose, sucrose, glycerol, mannitol, sorbitol) required to maintain an osmotic balance between the buffer and the interior of cells, should also be taken into account as this affects successful development of cells after treatment. For example, osmoticum for tissue electroporation of barley embryos was optimized by addition of mannitol to EB at concentrations between 300–700 mM [which equals 620–1100 mOsm $(kgH_2O)^{-1}$] and strongly affected callus initiation. Addition of polyethylene glycol (PEG) into EB was also tested in protoplast electroporation (Shillito et al. 1985) and increased the number of transformants, particularly when added after DNA at a ratio of 13% (w/v). Similarly, the presence of PEG at higher concentrations, 20–25% (v/v) in EB enhanced the transient GUS expression in wheat tissue electroporation (Zaghmout 1993). PEG which was also used in a 15% (w/v) concentration was found to be essential for detectable transgene expression in barley scutella (Hansch et al. 1996). However, this substance is not preferred for general use due to its toxic effects and careful removal is necessary by subsequent steps of washing. Carrier DNA is occasionally added to electroporation buffer to provide alternative substrates for any potential nucleases, e.g., calf thymus DNA at a concentration of 50 μg/ml (Xu and Li 1994). Transformation vectors for electroporation include closed circular or linearized plasmids containing selectable (e.g., *npt* II, *bar*) or reporter markers (e.g., *uidA*, CAT, *luc*) under the control of constitutive, inducible or organ-specific promoter sequences.

Regeneration into whole plants from transformed cell lines after electroporation can be achieved by optimizing tissue culture procedures for each cultivar. Size and age of explants (e.g., immature embryo, inflorescences), media composition and genotype are important factors for barley regeneration (Karp and Lazzeri 1992). Addition of several phytohormones such as 2,4-dichlorophenoxyacetic acid and cytokinins (e.g., benzylaminopurine) facilitates the regeneration process.

5.2.2 Electrical Variables

Two types of wave forms have been defined in electroporation: (1) an exponential decay pulse which produces a field that quickly rises to full strength, then gradually declines and (2) a square wave form which quickly rises to the desired amplitude, remains at that level and instantly declines. Commonly used commercial electroporators generate the former type of wave form. In this kind of device, the pulses released by the capacitor unit are characterized by two parameters: electrical field strength (kV/cm) and the time constant (τ:ms). Electrical field strength determines cell viability as well as gene transfer efficiency and can be controlled by setting the voltage (kV) directly in the electroporator unit. The time constant (τ) is determined by the resistance (R: ohms) and capacitance (C:μFd) in the complete circuit ($\tau = R \times C$). The capacitance (C) value can be selected on the capacitance extender unit within the electroporation device. The resistance of the sample depends on both its conductivity and the cross-sectional area of the path which can be adjusted by the use of various cuvette sizes (Gene Pulser BioRad, Operating Instructions and Application Guide).

In protoplast electroporation, the intensity of the electrical pulse required for efficient gene transfer is inversely proportional to the diameter of the protoplasts which varies from one plant species to another (Zimmerman et al. 1984). Two different approaches were used in protoplast electroporation: high voltage and short duration, e.g., 1.5 kV for 10 μs (Shillito et al. 1985) or lower voltage with longer duration, e.g., 350 V/cm for 54 ms (Fromm et al. 1985). It was suggested that short electrical pulses stimulate cell wall regeneration and plating efficiency in some plant protoplasts (Rech et al. 1987); however, these results could not be confirmed when different physical parameters were tested in barley (Mordhorst and Lörz 1992).

In intact tissue electroporation, the electrical field strength that produced optimal transient expressions varied for cereals in the range of values such as 275 V/cm for wheat scutellum cells (Klöti et al. 1993); 1000 V/cm for wheat embryoids (Gustafson et al. 1995); 900 V/cm for immature embryos of rice (Rao 1995); 375 V/cm for immature embryos of maize (D'Halluin et al. 1992) and 950 V/cm for immature scutella of barley (Hansch et al. 1996). Zaghmout (1993) also obtained increased GUS activity in slow growing embryogenic callus of wheat between 2000–6000 V/cm and suggested that more pores were generated at higher voltages. However, results from different experiments could not be compared realistically because different types of electroporators, tissue explants and genotypes were used in different studies.

5.3 Culture and Electroporation of Barley Explants

5.3.1 Protoplasts

The use of protoplasts in electroporation was preferred because the plant cell wall is a limiting factor for gene transfer and individual transformants can be obtained from protoplasts. In contrast, tissue electroporation may cause production of chimeric plants. Barley protoplasts were first isolated and cultured from fast-growing callus cultures (Koblitz 1976) and mesophyll tissues (Mesencev et al. 1976). In the following studies, the use of embryogenic cell suspensions and callus cultures were found to be ideal sources for regenerative protoplasts, however, regeneration was highly genotype-dependent (Karp and Lazzeri 1992). Mainly, two approaches were used for initiation of embryogenic suspension cultures for regenerable protoplasts in barley: culture of immature embryos directly in liquid medium (Yan et al. 1990) and the use of microspore-derived cultures as source material (Jahne et al. 1991b). A third, less common source was fertilized egg cells released from dissected ovules from which viable and dividing protoplasts were isolated and regenerated into whole plants by forming embryo-like structures (Holm et al. 1994). So far, regeneration from barley protoplasts resulted in the production of albino plantlets (Lührs and Lörz 1988), green plantlets (Lazzeri and Lörz 1990; Yan et al. 1990) and green fertile plants (Jahne et al. 1991a; Salmenkallio-Marttila 1994; Holm et al. 1994). The composition of protoplast culture media is based on standard formulations (MS, N_6, B_5, KM), but modified versions of this also provide sustained division and high yields of regeneration. Mainly, increased calcium ion concentration was found to be necessary for membrane stability, whereas reducing the ammonium ion concentration helps protoplast survival. Both auxin and cytokinins are needed in media for initial protoplast division (Nagata and Takebe 1970) and 2,4-dichlorophenoxyacetic acid is the most used auxin in producing regenerable cultures in barley. The type of carbon source (glucose, sucrose, maltose) and addition of several plant growth substances (coconut milk, casein hydrolysate, amino acids, amino acid derivatives, yeast and malt extracts) to the culture medium are also important for improving protoplast yield. For barley, L1 medium (Lazzeri et al. 1991) was efficiently used to culture protoplasts derived from cell suspensions (Jahne et al. 1991a; Funatsuki et al. 1992; Salmenkallio-Marttila 1994). For regeneration from protoplasts, lowering of the sugar concentration of the medium and addition of plant growth regulators (e.g., benzylaminopurine) which promote differentiation of the pro-embryos are essential. Nonmetabolizable osmoticum (sorbitol, mannitol) may be required in regeneration medium and a mixture of several carbon sources usually provides optimum results. In addition, several culture techniques improved the plating efficiencies in barley, such as using conditioned medium (Jorgensen et al. 1992), embedding protoplasts in alginate (or gelrite; Eigel and Koop 1989; Yan et al. 1990; Salmenkallio-Marttila 1994) or the use of feeder cells (Funatsuki et al. 1992; Salmenkallio-Marttila 1994).

Electroporation of barley protoplasts was first reported by Teeri et al. (1989). In this study, organelle transformation was demonstrated through the activity of *npt II* marker gene in mesophyll protoplasts. An extensive study regarding the parameters of protoplast electroporation of a two-rowed spring cultivar, Kymppi is presented in Table 5.1 (Salmenkallio-Marttila 1994). In part of this study, protoplasts were isolated according to Lazzeri et al. (1991) from embryogenic suspension cultures initiated from immature embryo and anther-derived calli. Isolated protoplasts were suspended in two different (chloride- and organic acid-based) electroporation buffers at a density of $3–5 \times 10^6\,ml^{-1}$ and 300 µl of this suspension was mixed with 30 µg of plasmid DNA (containing *npt II* driven by CaMV 35S promoter), chilled on ice for 10 min and subsequently electroporated. The protoplast suspension was kept on ice for 10 min after electroporation, mixed with culture medium and plated on regeneration medium. For electroporation, a range of 100–800 V/cm field strength with a constant capacitance of 200 µFd (single or two pulses) was used and 800 V/cm (single pulse) gave the highest transient expression of *npt II* (Table 5.1). However, this value reduced the plating efficiency by 77%. Moreover, 200–

Table 5.1. Optimization of the electroporation parameters for transformation of barley (*Hordeum vulgare* L. cv. Kymppi) protoplasts (Salmenkallio-Marttila 1994)

Parameters	Tested values	Optimal values
Field strength	100–800 V/cm (single pulse) 400–600 V/cm (two pulses)	800 V/cm
Capacitance	100, 200, 300, 400 μFd	200–300 μFd
DNA concentration	50–200 μg/ml	100–150 μg/ml
Promoter effect	CaMV 35S and maize *Adh I*	Maize *Adh I*
Electroporation buffer	Buffers based on KCl or organic acids	Organic acid-based buffer
Heat shock treatment	Electroporation with or w/o heat shock treatment	Heat shock-enhanced plating efficiency threefold

300 μFd capacitance was found to be optimal for higher transient expression of *npt II*. The time course of expression of the *uidA* gene was also examined in the same work by determining the number of cells expressing the transgene in protoplast cultures incubated for 24, 48 and 72 h after electroporation. Maximum GUS activity was detected 24 h after electroporation.

In addition, the embryogenic suspension cultures of Kymppi, 3–4 week old microspore cultures were also used as a protoplast source for electroporation (Salmenkallio-Marttila et al. 1995). Forty-two green plants were recovered without selection from 16.5×10^6 protoplasts electroporated by an optimized combination of 670 V/cm, 200 μFd. Regenerated plants were screened for transgene expression by gel assay and three were *npt II* transformants. Southern blots with DNA isolated from leaves of two parent plants, as well as the progeny of these plants, showed that the *npt II* gene was integrated as several copies in the barley genome. When inheritance of *npt II* was examined in T_2 progeny, Southern analysis also indicated that there were less than ten copies of inserted plasmid DNA with intact *npt II* and at least three copies of plasmid in which *npt II* disrupted during integration. All molecular data based on hybridization patterns confirmed that the parent plants originated from the same transformed cell.

5.3.2 Microspores

Microspore cultures are ideal recipients as haploid cells for several gene transfer techniques including microinjection, *Agrobacterium*-mediated transformation, biolistics and electroporation. Barley microspores were isolated by mechanical methods (e.g., gentle maceration with a Teflon rod) from anther cultures (Hoekstra et al. 1992) or whole spikes (Salmenkallio-Marttila 1994). Two types of microspores are characterized in Igri cultivar; the first is small, often plasmolysed, 35–40 μm in diameter and the second is large with granular cytoplasm, 40–60 μm in diameter. Only the latter types with mid-late to late uninucleate stage divided and grew in Igri cultivar and regenerated into green plants at a ratio of 50% under optimized conditions (Hoekstra et al. 1992). In the same study, duration of mannitol pre-treatment of donor anther cultures (4 days) and oxygen supply of both microspore and anther cultures influenced the regeneration. Specifically, mechanical isolation instead of natural shedding (dehiscing from anthers) yielded 6.5-fold more microspores in Igri. In another study, a mean of 1.1×10^5 microspores were isolated per spike of cultivar Kymppi (Salmenkallio-Marttila 1994). The dehydration treatment and increasing the maltose concentrations (e.g., 0.325 M) in both isolation buffer and growth medium improved regeneration plant production. Consequently, regeneration of barley microspores highly depends on the genotype and the most responsive barley cultivars are Igri, Kymppi and Gipel (Mordhorst and Lörz 1992). In well-optimized tissue culture systems, fertile barley plants could be produced by a mean of 169 green plants per spike (Salmenkallio-Marttila 1994). However, survival of microspores after electroporation was lower than that of protoplasts and intact tissues. The viability of electropermeabilized Igri microspores was examined by using an indicative dye, propidium iodide (PI; Joersbo et al. 1990). Microspores were isolated in late uninuclear stage and 6×10^4 microspores/ml were electroporated in the presence of PI solution after the viability was improved by a Percoll gradient centrifugation. In this study, optimal electrical conditions were five pulses of 100–400 μs at 120 V/mm in a commercial electroporator (Kruess, FRG, TA 750). Viability and microcalli formation were negatively affected by increasing electrical field strengths, as a result, only 5–17% of microspores survived after electropermeabilization and regenerated to microcalli, proembryos and plants.

5.3.3 Intact Tissues

The first report describing tissue electroporation of barley was that of Hansch et al. (1996). In this work, scutella of immature embryos were used for electroporation considering the highly competent nature of this tissue type for somatic embryogenesis. Isolated scutella of barley cultivars *H. vulgare* L. cvs. Golden Promise and Delita were electroporated in the presence of pVA13 [containing *luc* (firefly luciferase) and *pat* (phosphinothricin-*N*-acetyltransferase)] and pActI-F (containing gusA) plasmids. Between freshly isolated and 1-day pre-cultured scutella, the latter succeed in forming embryogenic callus after enzyme treatment, 2-h pre-incubation and electroporation. In this study, 300–1250 V/cm field strength within a single pulse and 17–8 ms pulse duration were tested using a commercial electroporator (BTX ECM 600 system) which produces an exponential decay wave form. Based on the transient expression of the luciferase gene and tissue viability, the optimal combination of field strength and pulse duration was chosen as 950 V/cm and 56 ms, respectively. When the form of plasmid DNA was tested with respect to transformation frequency, only linearized plasmid DNA resulted in detectable *luc* and GUS activity. Even though no molecular evidence of transgenes was presented in this work, both the potential use of tissue electroporation in barley and similar experimental results obtained in two genotypes were demonstrated.

Surface sterilization of seeds
(by incubation in 20% bleach for 20min and washing with sterile dH2O 3 times)
Removing seed coat and isolation of mature embryos, immersion in 99% ethanol for 5sec, washing with sterile dH2O twice, transfer to CULTURE MEDIUM[a] (CM: Murashige and Skoog (1962) salts, 30gr/l sucrose, 2mg/l 2,4-D and 0.8%agar)
Electroporation 1, 2 and 3 days after
Transfer embryos into a petri dish containing Buffer A
(20mM CaCl2, 10mM Hepes, 0.4M mannitol pH 5.6)
Incubation for 1h at room temperature
Picking up embryos to a new petri dish
containing Buffer A plus 0.3% macerase and 0.5% cellulase
Incubation for 15 min at room temperature
Washing embryos with sterile dH2O twice,
transfer 10-12 embryos to each ice-cold electroporation cuvettes[b]
containing Buffer A plus 50mM KCl
Addition of circular plasmid DNA (pBI 121) (total 30μg) into each cuvettes
Incubation on ice for 30min
Electroporation (500V/cm-500μFd, single pulse) by Gene Pulser (BioRad)
Transfer of embryos to REGENERATION MEDIUM
[RM:CM plus 0.1mg/l of benzylamino purine (BAP)]
Culture on regeneration medium for 3 days
[Assay of GUS activity on cultured explants]
Transfer of embryos to SELECTIVE MEDIUM (SM:RM plus 40mg/l Geneticin)
3 months
DNA isolation from plant leaves and Southern blot analysis
[Assay of GUS activity in 3 month-old plant parts]

Another explant source for tissue electroporation is mature embryos from seeds (Gürel and Gözükırmızı 2000). We developed a procedure (Fig. 5.1),

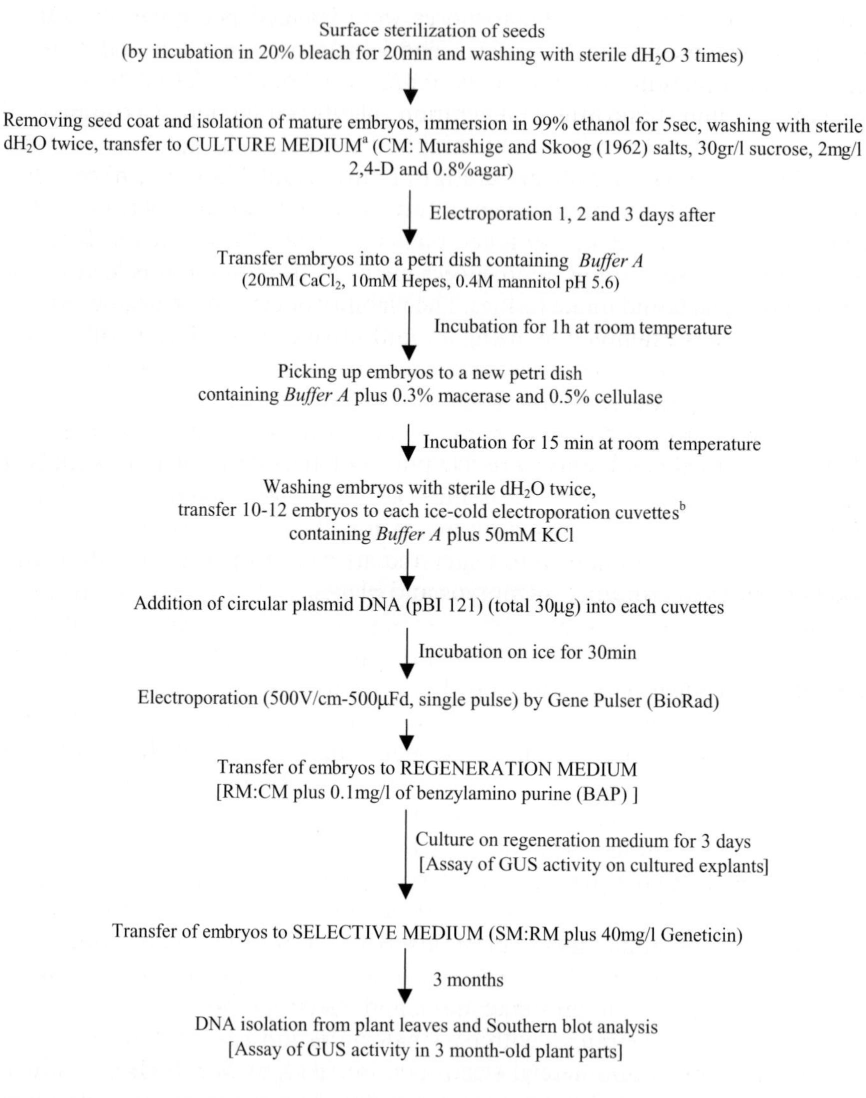

[a]All the cultures kept in a growth chamber at 25°C under a photoperiod of 8/16h (light/dark)
[b]Gene Pulser (BioRad) cuvettes with 0.4cm gap distance

Fig. 5.1. Experimental steps for tissue electroporation of intact mature embryos (*H. vulgare* L. cv. Zafer 160)

mainly modified from that of D'Halluin et al. (1992) for transformation of barley embryos in our laboratory. The material used in the study is a Turkish cultivar Zafer 160, which is a six-rowed (2n = 2x = 14) winter-type barley. After the surface sterilization of dry seeds by incubation in 20% bleach for 20 min and subsequent washing with sterile distilled water three times, embryos were isolated and cultured on MS (Murashige and Skoog 1962) medium supplemented with 30 g/l sucrose, 2 mg/l 2,4-D and 0.8% agar. Approximately 20–25 embryos were placed on Petri dishes and incubated in a growth chamber at 25°C under a photoperiod of 8/16 h (light/dark). Embryos in different culture periods (1, 2 and 3 days after culture) were electroporated using a commercial electroporator (BioRad). Before electroporating the explants, they were incubated in a filter-sterilized plasmolysis buffer (designated as Buffer A) for 1 h at room temperature and transferred to enzyme solution for another 15 min. After washing with sterile distilled water, 10–12 embryos were carefully transferred to ice-cold electroporation cuvettes containing Buffer A plus 50 mM KCl. The plasmid we used was pBI 121 which contains GUS (*uidA*; *E. coli* β-glucuronidase) gene driven by the 35S CaMV promoter and *nptII* (neomycin phosphotransferase II) driven by nopaline synthetase (NOS) promoter. Thirty μg of circular plasmid DNA was added to each cuvette and electroporation was applied under different voltage-capacitance values within a single pulse. We mainly investigated the effect of voltage-capacitance values on gene transfer and germination capacity. We paid special attention to the pre-culture period before electroporation and selective agent concentration (Geneticin; G418) which had to be adjusted for eliminating untransformed cells. Although the germination frequency of nonelectroporated embryos was nearly 100% in our study, electroporation (on the second day of culture) by different combinations of three electrical field strength (500, 750, 1000 V/cm) and two capacitance values (500 and 960 μFd) resulted in a different profile of germination. Germination frequencies of mature embryos were; 75% in 500 V/cm–500 μFd; 69% in 500 V/cm–960 μFd; 48% in 750 V/cm–500 μFd; 22.5% in 750 V/cm–960 μFd; 10% in 1000 V/cm–500 μFd and 48.5% in 1000 V/cm–960 μFd. As a result, a voltage of 0.2 kV (500 V/cm) and a capacitance of 500 μFd represented the optimal combination, yielding germination frequencies of up to 75% in our study. Furthermore, we determined the pre-culture period that produces embryos more resistant to the damage following electroporation. Table 5.2 shows that longer periods of culture protected the embryos from the damage caused by electroporation procedure. Specifically, electroporation performed on the third day of culture resulted in a substantially higher germination frequency (95%) than that applied on the second and first culture days (24 and 20%, respectively; see Table 5.2, second column).

Furthermore, embryos electroporated on the first day of culture failed completely to germinate on the geneticin-selective medium. An additional day of culture allowed germination at lower levels of the antibiotic (40 and 80 mg/l), but still did not permit germination at a higher concentration of geneticin (100

Table 5.2. Effects of different culture periods before electroporation and antibiotic concentrations used in selection on the rate of germination. Barley embryos electroporated using 500 V/cm voltage and 500 μFd capacitance

Culture period	Geneticin concentration (mg/l)			
	0	40	80	100
1 day	20 (8/40)[a]	0.0 (0/32)	0.0 (0/28)	0.0 (0/22)
2 day	24.1 (13/54)	10.0 (3/30)	5.0 (2/40)	0.0 (0/42)
3 day	95.0 (38/40)	12.5 (5/40)	5.6 (2/36)	3.3 (1/30)

[a]The values in parenthesis indicate the number of germinated plants per total number of electroporated embryos in an individual experiment.

mg/l; Table 5.2). Finally, Table 5.2 shows that 3 days of culture allowed the electroporated embryos to survive in the highest concentration of geneticin (100 mg/l). These results suggest that the germination capacity of embryos cultured on both MS and selective media depends on the number of actively dividing cells within the embryos, which are known to increase with the duration of the culture period. Higher concentrations of geneticin also reduced the germination and partially caused albino phenotypes. The highest percentage of embryos exhibiting the cell clusters and showing GUS expression in our study was 40% and those were electroporated with the 500 V/cm and 960 μFd combination (Fig. 5.2A). GUS-expressing cell clusters were also observed in mature leaves of 3-month-old plantlets derived from those embryos (Fig. 5.2B). Finally, our electroporation and regeneration conditions using selection in the presence of 40 mg/l of geneticin allowed both shoot and root formation in the transgenic plantlets (Fig. 5.2C).

The use of mature embryos for electroporation was also tested in rice (Xu and Li 1994). Half of the embryos were electroporated with a construct carrying *npt II* gene driven by NOS promoter and regenerated plants were produced through the geneticin-resistant calli. Of the 35 R_0 plants that were transferred to soil and grew to maturity, 23 expressed *npt II* gene. Inheritance of *npt II* gene was then confirmed by molecular analysis starting from callus lines, R_0 and R_1 plants.

These works in rice and barley demonstrate the feasibility of electroporation procedures for mature and immature embryos. Moreover, the use of mature embryos from seeds instead of other explants eliminates the initial steps before tissue culture (growing plants in a greenhouse) and seasonal limitations, thus enabling one to make extensive optimizations with a high number of samples.

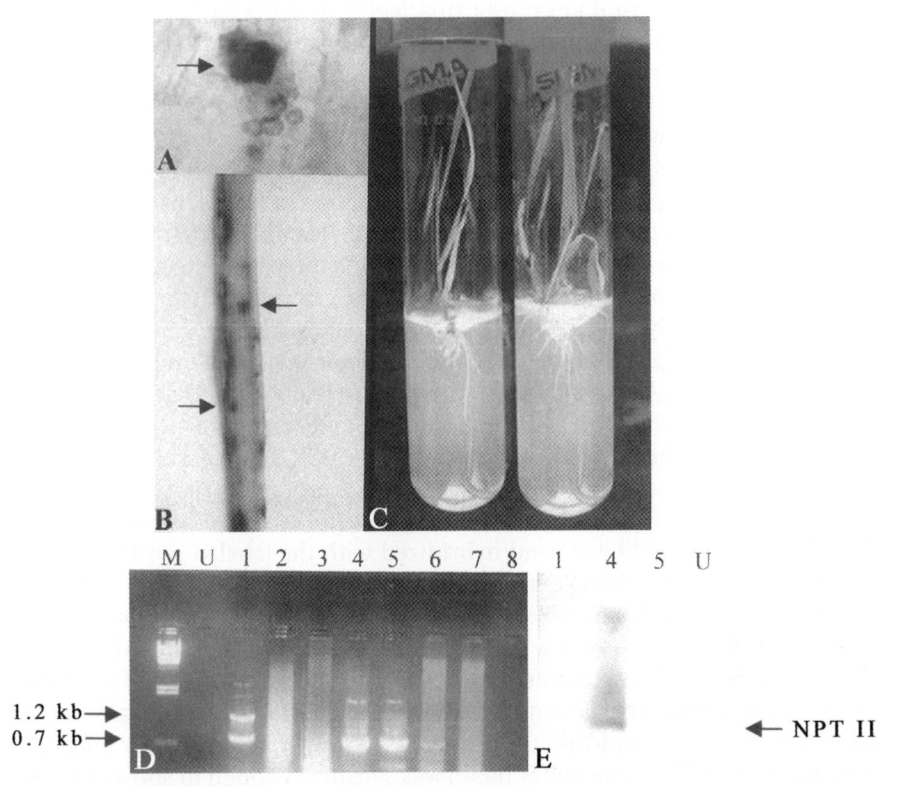

Fig. 5.2. **A** GUS gene expression in a transversal section of a mature embryo electroporated at 500 V/cm voltage and 960 μFd capacitance. Magnification: 100×. **B** GUS-expressing cell clusters in a mature leaf segment from a 3-month-old plantlet grown on MS+BAP+G418 medium. **C** Regeneration and root formation from plantlets (*lanes 4* and *5* in **D**). **D** PCR analysis of eight barley plants grown in the presence of 40 mg/l G418. *Lane 1* PCR products corresponding to the *npt II* (0.7 kb) and GUS (1.2 kb) genes; *lanes 1, 4, 5 npt II*-positive plantlets, *lane U* untransformed plant as a negative control, *lane M* molecular size markers (λ-DNA digested with *Hin*dIII). **E** Southern blot analysis of DNA from transformants (*lanes 1, 4, 5*) and untransformed control (*U*) cut with *Hin*dIII and probed with labeled *npt II* cDNA

5.3.3.1 Analysis and Inheritance of Transgenes in Electroporated Tissues

GUS gene expression can be histochemically assayed by Jefferson (1987) incubation of initial explants and other plant sections in a staining solution (1.5 mM potassium ferrocyanide, 1.5 mM potassium ferricyanide, 0.9% Triton X-100, 0.3% 5 bromo-4 chloro-3 indolyl β-D-glucuronic acid (X-gluc), 20% methanol) overnight at 37°C. GUS activity can also be visualized microscopically after transversal sections of embryos are excised under aseptic conditions after electroporation (Fig. 5.2A). In our study, chimeric expression of GUS was detected on primary root apex, root tips, leaf primordia and mature leaf segments from 3-month-old plantlets (Fig. 5.2B). In five plantlets derived from electroporated embryos grown on 80–100 mg/l geneticin (Table 5.2, columns 4 and 5), neither GUS nor *npt II* genes was detected by PCR. In contrast, of eight of the plantlets grown on 40 mg/l of geneticin (Table 5.2, column 3), one plant contained both GUS and *npt II* genes and two plants contained only *npt II* gene (Fig. 5.2D: lanes1, 4, 5). GUS gene was not amplified in two plants even though both genes were present on the same pBI 121 plasmid. This situation is not unique as similar results have been reported for the transformation of barley protoplasts with the *npt II* and GUS genes (Lazzeri et al. 1991). Using Southern blot analysis, Lazzeri et al. (1991) showed that some geneticin-resistant lines completely lacked the GUS gene or contained only parts of its sequence. Plant DNAs isolated from the plants (Walbot 1988) shown in lanes 1, 4 and 5 of Fig. 5.2D were also subjected to Southern blot analysis (Dwivedi et al. 1994) to further confirm the results from PCR. A DNA fragment corresponding to a partial *npt II* gene was used as a probe. When genomic DNAs were digested with *Hin*dIII (cleaves once between the nos terminator and 35S CaMV promoter of pBI 121) and hybridized with the labeled *npt II* gene, one hybridizing fragment was detected in one line that had been previously identified as *npt II*-positive by PCR (Fig. 5.2E). The other two plants tested, however, did not produce any fragment. This could be possibly due to the chimeric nature of these plants with respect to the target (*npt II*): DNA that exists in transformed tissues and was diluted by the DNA from untransformed tissues to such an extent that this target is no longer detectable by Southern hybridization. PCR, on the other hand, was sensitive enough to show the presence of the *npt II* gene in the DNA isolated from all three plants. Inheritance of the transgenes in progeny of transformed plants could not be examined mainly due to difficulties in obtaining seeds from regenerated/transformed plants in our work.

Recently, tissue electroporation has been tested using inflorescences of tritordeum plants which are barley – wheat hybrids (*Hordeum chilense* Roem. × *Triticum turgidum* L. Conv. *durum*; He et al. 2001). A total of 361 explants (from Line HT174) were electroporated under 550 V/cm–960 μFd with pAHC25 plasmid carrying *uidA* and *bar* genes both driven by maize ubiquitin promoter. Transgene expressions of GUS and *bar* were assayed according to Barro et al. (1998) on leaf tissues of T_0 and progeny plants and applying BASTA solution to leaves, respectively. As a result, two transgenic lines (designated as P7 and P8) expressing *uidA* grew to maturity and set seeds. PCR analysis showed that both *bar* and *uidA* genes were present in T_1 progeny. In addition, extracted DNA from plants was digested with *Hin*dIII (releases a 4.2-kb fragment containing ubi promoter and *uidA* gene) and *Sac*I (cleaves once within plasmid pAHC25) and probed with *uidA* coding region. Particularly in *Hin*dIII digestion, identical banding patterns in the expected size (4.2 kb) were observed in all tested lines of T_1 and T_2. Overall data indicated that two lines (P7, P8) originated from the same transformation event.

5.4 Conclusions and Future Perspectives

Electroporation-mediated transformation of barley has been successful using cultured protoplasts, microspores, immature embryo-derived scutellar tissues and mature embryos. Regenerative tissue culture procedures from both microspores and protoplasts are genotype-dependent and labor-intensive. At this point, an understanding of the genetic and developmental basis of in vitro cultures will be helpful to overcome regeneration-related problems in barley (Lemaux et al. 1999). Moreover, physical conditions during electroporation highly affect cell viability. Three fertile transgenic plants expressing the *npt II* gene were recovered in Kymppi cultivar (Salmenkallio-Marttila 1994) by a procedure which takes 30–34 weeks from collection of spikes for microspore isolation to harvest of seeds. The long time period of tissue culture in this procedure may be a serious problem as this may cause increased rates of somaclonal variations.

Tissue electroporation, on the other hand, seems to be more promising as a feasible procedure for barley transformation. Recent production of fertile transgenic wheat (Sorokin et al. 2000) and wheat – barley hybrid plants (He et al. 2001) by intact tissue electroporation supports this idea. In addition to the scutellar tissues and mature embryos, other tissues such as inflorescences, calli and apical meristems can also be tested as potential targets for electroporation. The culture of these explants and gene transfer procedure will probably take less time than that of the cell-based systems. Specifically, use of mature embryos have advantages such as fast regeneration through germination, genotype-independency and abundancy in every season.

The transformed plant is expected to contain a single copy of the transgene which will be passed on in a Mendelian fashion to its progeny. However, transgene integrations are found to be quite complex in most transformation events. Therefore, an understanding of transgene integration in higher plants and efforts for higher expressions of genes such as by scaffold attachment regions (Spiker and Thompson 1996) will be helpful in designing new vectors for electroporation. In addition, transgenes introduced by *Agrobacterium*-mediated transformation as well as DNA fragments introduced by direct gene transfer are prone to silencing. Critical for the triggering of the silencing phenomenon is the presence of homologous sequences residing within two or more introduced transgenes or within an introduced transgene and a homologous endogenous gene. Transgene silencing is more commonly observed in plants with multiple (linked or unlinked) copies than in plants with a single copy of the transgene (Linn et al. 1990). Therefore, a large number of explants in a transformation experiment is important to find the desired genotypes among the large numbers of variants (Hansen and Wright 1999). In this respect, tissue electroporation procedures described in this chapter are convenient and economical for production of transgenic barley cultivars in high frequency for future applications.

Acknowledgements. This study was supported in part by NATO-TU-Biotechnology grant No. 842 and a Research Foundation of Istanbul University grant No. 1370/081299.

References

Ahloowalia BS (1987) Plant regeneration from embryo-callus culture in barley. Euphytica 36: 659–665

Ahokas H (1989) Transfection of germinating barley seed electrophoretically with exogenous DNA. Theor Appl Genet 77:469–472

Barro F, Cannell ME, Lazzeri PA, Barcelo P (1998) The influence of auxins on transformation of wheat and tritordeum and analysis of transgene integration patterns in transformants. Theor Appl Genet 97:684–695

Becker J, Heun M (1995) Barley microsatellites: allele variation and mapping. Plant Mol Biol 27(4):835–845

Bennet MD, Smith LB (1976) Nuclear DNA amounts in angiosperms. Philos Trans R Soc Lond B Biol Sci 274:227–274

Bolik M, Koop HU (1991) Identification of embryogenic microspores of barley (*Hordeum vulgare* L.) by individual selection and culture and their potential for transformation by microinjection. Protoplasma 162:61–68

Breiman A (1985) Plant regeneration from *Hordeum spontaneum* and *Hordeum bulbosum* immature embryo-derived calli. Plant Cell Rep 4:70–73

Cheng TY, Smith HH (1975) Organogenesis from callus culture of *Hordeum vulgare*. Planta 123: 307–310

Cheng M, Fry JE, Pang S, Zhou H, Hironaka CM, Duncan DR, Conner TW, Wan Y (1997) Genetic transformation of wheat mediated by *Agrobacterium tumefaciens*. Plant Physiol 115:971–980

Chibbar RN, Kartha KK, Datla RSS, Leung N, Caswell K, Mallard CS, Steinhauer L (1993) The effect of different promoter-sequences on transient expression of *gus* reporter gene in cultured barley (*Hordeum vulgare* L.) cells. Plant Cell Rep 12:506–509

Dekeyser RA, Claes B, DeRycke RMU, Habets ME, Van Montagu MC, Caplan AB (1990) Transient gene expression in intact and organized rice tissues. Plant Cell 2:591–602

D'Halluin K, Bonne E, Bossut M, DeBeuckeleer M, Leemans J (1992) Transgenic maize plants by tissue electroporation. Plant Cell 4:1495–1505

Dunwell JM (1986) Barley. In: Evans DA, Sharp R, Ammirato PV (eds) Handbook of plant cell culture, vol 4. MacMillan, London, pp 339–369

Dwivedi UN, Capbell WH, Yu J, Datla RSS, Bugos RC, Chiang VL, Dodila GK (1994) Modification of lignin biosynthesis in transgenic *Nicotiana* through expression of an antisense *O-methyltransferase* gene from *Populus*. Plant Mol Biol 26:61–71

Eigel L, Koop HU (1989) Nurse culture of individual cells: regeneration of colonies from single protoplasts of *Nicotiana tabacum*, *Brassica napus* and *Hordeum vulgare* L. J Plant Physiol 134: 577–581

Funatsuki H, Lörz H, Lazzeri PA (1992) Use of feeder cells to improve barley protoplast culture and regeneration. Plant Sci 85:179–187

Funatsuki H, Kuroda M, Lazzeri PA, Müller E, Lörz H, Kishinami I (1995) Fertile transgenic barley generated by direct DNA transfer to protoplasts. Theor Appl Genet 91:707–712

Fromm M, Taylor LP, Walbot V (1985) Expression of genes transferred into monocot and dicot plant cells by electroporation. Proc Natl Acad Sci USA 82:5824–5828

Goldstein CS, Kronstrad WE (1986) Tissue culture and plant regeneration from immature embryos of barley *Hordeum vulgare*. Theor Appl Genet 71:631–636

Gözükırmızı N, Arı S, Oraler G, Okatan Y, Ünsal N (1990) Callus induction, plant regeneration and chromosomal variations in barley. Acta Bot Neerl 39:379–387

Gustafson VD, Baenziger PS, Mitra A, Kaeppler HF, Papa CM, Kaeppler SM (1995) Electroporation of wheat anther culture-derived embryoids. Cereal Res Comm 23:207–213

Gürel F, Gözükırmızı N (2000) Optimization of gene transfer into barley (*Hordeum vulgare* L.) mature embryos by tissue electroporation. Plant Cell Rep 19:787–791

Hagio T, Hirabayashi T, Machii H, Tomotsune H (1995) Production of fertile transgenic barley (*Hordeum vulgare* L.) plant using the hygromycin-resistance marker. Plant Cell Rep 14: 329–334

Hansch R, Koprek T, Heydemann H, Mendel RR, Schulze J (1996) Electroporation-mediated transient gene expression in isolated scutella of *Hordeum vulgare*. Physiol Plant 98:20–27

Hansen G, Wright MS (1999) Recent advances in the transformation of plants. Trends Plant Sci 4:226–231

Harwood WA, Bean SJ, Chen DF, Mullineaux PM, Snape JW (1995) Transformation studies in *Hordeum vulgare* using a highly regenerable microspore system. Euphytica 85:113–118

Hayes PM, Chen PQ, Kleinhofs A, Kilian A, Mather DE (1996) Barley genome mapping and its applications in methods of genome analysis in plants. In: Jauhar PP (ed) Methods of genome analysis in plants. CRC Press, Boca Raton, pp 229–249

He GY, Lazzeri PA, Cannell ME (2001) Fertile transgenic plants obtained from tritordeum inflorescences by tissue electroporation. Plant Cell Rep 20:67–72

Hoekstra S, Zijderveld MH, Louwerse JD, Heidekamp F, Mark F (1992) Anther and microspore culture of *Hordeum vulgare* L. cv. Igri. Plant Sci 86:89–96

Holm PB, Knudsen S, Mouritzen P, Negri D, Olsen FL, Roue C (1994) Regeneration of fertile barley plants from mechanically isolated protoplasts of the fertilized egg cell. Plant Cell 6:531–543

Holm PB, Olsen O, Schnorf M, Brinch-Pedersen H, Knudsen S (2000) Transformation of barley by microinjection into isolated zygote protoplasts. Transgenic Res 9:21–32

Huang B, Sunderland N (1982) Temperature-stress pretreatment in barley anther culture. Ann Bot 49:77–88

Jahne A, Lazzeri PA, Lörz H (1991a) Regeneration of fertile plants from protoplasts derived from embryogenic cell suspensions of barley (*Hordeum vulgare* L.) Plant Cell Rep 10:1–6

Jahne A, Lazzeri PA, Jager-Gussen M, Lörz H (1991b) Plant regeneration from embryogenic cell suspensions derived from anther cultures of barley (*Hordeum vulgare* L.). Theor Appl Genet 82:74–80

Jahne A, Becker D, Brettschneider R, Lörz H (1994) Regeneration of transgenic, microspore derived, fertile barley. Theor Appl Genet 89:525–533

Jefferson RA (1987) Assaying chimeric genes in plants: GUS gene fusion system. Plant Mol Cell Biol 4:347–357

Jensen LG, Olsen O, Kops O, Wolf N, Thomsen KK (1996) Transgenic barley expressing a protein-engineered, thermostable (1,3–1,4)-β-glucanase during germination. Proc Natl Acad Sci USA 93:3487–3491

Joersbo M, Jorgensen RB, Olesen P (1990) Transient electropermeabilization of barley (*Hordeum vulgare* L.) microspores to propidium iodide. Plant Cell Tissue Organ Cult 23:125–129

Jorgensen RB, Jensen CJ, Andersen B, Bothmer R (1986) High capacity of plant regeneration from callus of interspecific hybrids with cultivated barley (*Hordeum vulgare* L.). Plant Cell Tissue Organ Cult 6:199–207

Jorgensen RB, Andersen B, Andersen JM (1992) Effects and characterization of the conditioning medium that increase colony formation from barley (*Hordeum vulgare* L.) protoplasts. J Plant Physiol 140:328–333

Junker B, Zimmy J, Lührs R, Lörz H (1987) Transient expression of chimeric genes in dividing and nondividing cereal protoplasts after PEG induced DNA uptake. Plant Cell Rep 6:329–332

Kao KN, Horn DC (1982) A method for induction of pollen plants in barley In: Fujiwara A (ed) Plant tissue culture. Maruzen, Tokyo, pp 529–530

Kao KN, Saleem M, Abrams S, Petras S, Horn D, Mallard C (1991) Culture conditions for induction of green plants from barley microspores by anther culture methods. Plant Cell Rep 9: 595–601

Karp A, Steele SH, Breiman A, Shewry PR, Parmar S, Jones MGK (1987) Minimal variation in barley plants regenerated from cultured immature embryos. Genome 29:405–412

Karp A, Lazzeri PA (1992) Regeneration, stability and transformation of barley. In: Shewry PR (ed) Barley: genetics, biochemistry, molecular biology and biotechnology. Alden Press/CAB International, Oxford, pp 549–571

Kartha KK, Chibbar RN, Georges F, Leung N, Caswell K, Kendall E, Qureshi J (1989) Transient expression of chloramphenicol acetyltransferase (CAT) gene in barley cell cultures and immature embryos through microprojectile bombardment. Plant Cell Rep 8:429–432

Katoh Y, Hasegawa T, Suzuki T, Fuji T (1986) Plant regeneration from the callus derived from mature embryos of Hiproly barley, *Hordeum distichum* L., culture. Agric Biol Chem 50:761–762

Kihara M, Saeki K, Ito K (1998) Rapid production of fertile transgenic barley (*Hordeum vulgare* L.) by direct gene transfer to primary callus-derived protoplasts. Plant Cell Rep 17:937–960

Klöti A, Iglesias VA, Wünn J, Burkhardt PK, Datta SK, Potrykus I (1993) Gene transfer by electroporation into intact scutellum cells of wheat embryos. Plant Cell Rep 12:671–675

Knudsen S, Müller M (1991) Transformation of the developing barley endosperm by particle bombardment. Planta 185:330–336

Koblitz H (1976) Isolierung und Kultivierung von Protoplasten aus Calluskulturen der Gerste. Biochem Physiol Pflanz 170:287–293

Köhler F, Wenzel G (1985) Regeneration of isolated barley microspores in conditioned media and trials to characterize the responsible factor. J Plant Physiol 121:181–191

Koprek T, Haensch R, Nerlich A, Mendel RR, Schulze J (1996) Fertile transgenic barley of different cultivars obtained by adjustment of bombardment conditions to tissue response. Plant Sci 119:79–91

Kott LS, Kasha KJ (1984) Initiation and morphological development of somatic embryoids from barley cell cultures. Can J Bot 62:1245–1249

Laursen CM, Krzyzek RA, Flick CE, Anderson PC, Spencer TM (1994) Production of fertile transgenic maize by electroporation of suspension culture cells. Plant Mol Biol 24:51–61

Lazzeri PA, Lörz H (1990) Regenerable suspension and protoplast cultures of barley and stable transformation via DNA uptake into protoplasts. In: Lycett GW, Grierson D (eds) Genetic engineering of crop plants. Butterworths, London, pp 231–237

Lazzeri PA, Brettschneider R, Lührs R, Lörz H (1991) Stable transformation of barley via PEG-induced direct DNA uptake into protoplasts. Theor Appl Genet 81:437–444

Lee BT, Murdock K, Topping J, Kreis M, Jones MGK (1989) Transient gene expression in aleurone protoplasts isolated from developing caryopses of barley and wheat. Plant Mol Biol 13:21–29

Lemaux PG, Cho MJ, Zhang S, Bregitzer P (1999) Transgenic cereals: *Hordeum vulgare* L. (barley) In: Vasil IK (ed) Molecular improvement of cereal crops. Kluwer Academic, London, pp 255–316

Linn F, Heidmann I, Saedler H, Meyer P (1990) Epigenetic changes in the expression of the maize A1 gene in *Petunia hybrida*: role of numbers of integrated gene copies and state of methylation. Mol Gen Genet 222:329–336

Lührs R, Lörz H (1987) Plant regeneration in vitro from embryogenic cultures of spring- and winter-type barley (*Hordeum vulgare* L.). Theor Appl Genet 75:16–25

Lührs R, Lörz H (1988) Initiation of morphogenic cell suspensions and protoplast cultures of barley (*Hordeum vulgare* L). Planta 175:71–81

Lührs R, Nielsen K (1992) Microspore cultures as donor tissue for the initiation of embryogenic cell suspensions in barley. Plant Cell Tissue Organ Cult 31:169–178

Lupotto E (1984) Callus induction and plant regeneration from barley mature embryos. Ann Bot 54:523–529

Manninen OM (2000) Genetic mapping of important traits in barley breeding. Academic Dissertation, ISBN 951-45-9007-4. Division of Genetics, Department of Biosciences, University of Helsinki, February 2000

Mendel RR, Müller B, Schulze J, Kolesnikov V, Zelenin A (1989) Delivery of foreign genes to intact barley cells by high-velocity microprojectiles. Theor Appl Genet 78:31–34

Mendel RR, Clauss E, Hellmund R, Schulze J, Steinbiss HH, Tewes A (1990) Gene transfer to barley. In: Nijkamp HJJ, Plas LHW, Van Aartrijk J (eds) Progress in plant cellular and molecular biology. Kluwer, Dordrecht, pp 73–78

Mesencev AV, Butenko RG, Rodionova NA (1976) Obtention of isolated protoplasts from mesophyll of perennial crop plants and barley. Fiziol Rastenii 23:508–512

Michalek W, Weschke W, Pleissner KP, Graner A (2002) EST analysis in barley defines a unigene set comprising 4000 genes. Theor Appl Genet 104:97–103

Mordhorst AP, Lörz H (1992) Electrostimulated regeneration of plantlets from protoplasts derived from cell suspensions of barley (*Hordeum vulgare*). Physiol Plant 85:289–294

Müller B, Schulze J, Wegner U (1989) Establishment of barley cell suspension cultures of mesocotyl origin suitable for isolation of dividing protoplasts. Biochem Physiol Pfl 185:123–130

Murashige T, Skoog F (1962) A revised medium for rapid growth and bioassays with tobacco tissue cultures. Physiol Plant 15:473–497

Nagata T, Takebe I (1970) Cell wall regeneration and cell division in isolated tobacco mesophyll protoplasts. Planta 92:301–308

Neumann E, Schaefer-Ridder M, Wang Y, Hofschneider PH (1982) Gene transfer into mouse lyoma cells by electroporation in high electric fields. EMBO J 1:841–845

Nevo E (1992) Origin, evolution, population genetics and resources for breeding of wild barley, *Hordeum spontaneum*, in the Fertile Crescent. In: Shewry PR (ed) Barley: genetics, biochemistry, molecular biology and biotechnology. Alden Press/CAB International, Oxford, pp 19–43

Olsen FL (1991) Isolation and cultivation of embryogenic microspores from barley (*Hordeum vulgare* L.). Hereditas 115:255–266

Öztürk ZN, Talame V, Deyholos M, Michalowski CB, Galbraith DW, Gözükırmızı N, Tuberosa R, Bohnert HJ (2002) Monitoring large-scale changes in transcript abundance in drought and salt-stressed barley. Plant Mol Biol 48:551–573

Padua VL, Ferreira RP, Meneses L, Uchoa N, Margis-Pinherio M, Mansur E (2001) Transformation of Brazilian elite *indica*-type rice (*Oryza sativa* L.) by electroporation of shoot apex explants. Plant Mol Biol Rep 19:55–64

Patel M, Johnson JS, Brettell RIS, Jacobsen J, Xue GP (2000) Transgenic barley expressing a fungal xylanase gene in the endosperm of the developing grains. Mol Breed 6:113–123

Piccirilli M, Arcioni S (1991) Haploid plants regenerated via anther culture in wild barley (*Hordeum spontaneum* C. Kock). Plant Cell Rep 10:273–276

Ramsay L, Macaulay M, Degli-Ivanissevich S, MacLean K, Cardle L, Fuller J, Edwards KJ, Juvesson S, Morgante M, Massari A, Maestri E, Marmiroli N, Sjakste T, Ganal M, Powell W, Waugh R (2000) A simple sequence repeated-based linkage map of barley. Genetics 156:1997–2005

Rao KV(1995) Transient gene expression in electroporated immature embryos of rice (*Oryza sativa* L.). J Plant Physiol 147:71–74

Rech EL, Ochatt SJ, Chand PK, Power JB, Davey MR(1987) Electro-enhancement of division of plant protoplast-derived cells. Protoplasma 141:169–176

Rengel Z (1987) Embryogenic callus induction and plant regeneration from cultured *Hordeum vulgare* mature embryos. Plant Physiol Biochem 25:43–48

Ritala A, Aikasalo R, Aspegren K, Salmenkallio-Marttila M, Akerman S, Mannonen L, Kurten U, Puupponen-Pimia R, Teeri TH, Kauppinen V (1995) Transgenic barley by particle bombardment: inheritance of the transferred gene and characteristics of transgenic barley plants. Euphytica 85:81–88

Rogers SW, Rogers JC (1992) The importance of DNA methylation for stability of foreign DNA in barley. Plant Mol Biol 18:945–961

Rotem-Abarbanell D, Breiman A (1989) Plant regeneration from immature and mature embryo derived calli of *Hordeum marinum*. Plant Cell Tissue Organ Cult 16:207–216

Salmenkallio-Martilla M (1994) Regeneration of fertile barley plants from protoplasts and production of transgenic barley by electroporation. Academic Dissertation, ISBN 951-38-4640-7 (VTT Publications). Plant Physiology Division, Department of Botany, Faculty of Science, University of Helsinki, Nov 1994

Salmenkallio-Marttila M, Aspegren K, Akerman S, Kurten U, Mannonen L, Ritala A, Teeri TH, Kauppinen V (1995) Transgenic barley (*Hordeum vulgare* L.) by electroporation of protoplasts. Plant Cell Rep 15:301–304

Seguin-Swartz G, Kott L, Kasha KJ (1984) Development of haploid cell lines from immature barley, *Hordeum vulgare* embryos. Plant Cell Rep 3:95–97

Shewry PR (1992) Barley: genetics, biochemistry, molecular biology and biochemistry. Alden Press/CAB International, Oxford

Shillito RD, Saul MW, Paszkowski J, Müller M, Potrykus I (1985) High efficiency direct gene transfer to plants. Biotechnology 3:1099–1105

Songstad DD, Halaka FG, DeBoer DL, Armstrong CL, Hinchee MAW, Ford-Santino CG, Brown SM, From ME, Horsch RB (1993) Transient expression of GUS and anthocyanin constructs in intact maize immature embryos following electroporation. Plant Cell Tissue Organ Cult 33: 195–201

Sorokin AP, Ke XY, Chen DF, Elliott MC (2000) Production of fertile transgenic wheat plants via tissue electroporation. Plant Sci 156:227–233

Spiker S, Thompson WF (1996) Nuclear matrix attachment regions and transgene expression in plants. Plant Physiol 110:15–21

Sreenivasulu N, Altschmied L, Panitz R, Hahnel U, Michalek W, Weschke W, Wobus U (2002) Identification of genes specifically expressed in maternal and filial tissues of barley caryopses: a cDNA array analysis. Mol Genet Genomics 266:758–767

Stiff CM, Kilian A, Zhou H, Kudrna DA, Kleinhofs A (1995) Stable transformation of barley callus using biolistic particle bombardment and the phosphinothricin acetyltransferase (*bar*) gene. Plant Cell Tissue Organ Cult 40:243–248

Tada Y, Sakamoto M, Fujimura T (1990) Efficient gene introduction into rice by electroporation and analysis of transgenic plants: use of electroporation buffer lacking chloride ions. Theor Appl Genet 80:475–480

Teeri TH, Patel GH, Aspegren A, Kauppinen V (1989) Chloroplast targeting of neomycin phosphotransferase II with a pea transit peptide in electroporated barley mesophyll protoplasts. Plant Cell Rep 8:187–190

Thomas MR, Scott KJ (1985) Plant regeneration by somatic embryogenesis from callus initiated from immature embryos and immature inflorescences of *Hordeum vulgare*. J Plant Physiol 12: 159–169

Tingay S, McElroy D, Kalla R, Fieg S, Wang M, Thornton S, Brettell R (1997) *Agrobacterium tumefaciens*-mediated barley transformation. Plant J 11:1369–1376

Töpfer R, Gronenborn B, Schell J, Steinbiss HH (1989) Uptake and transient expression of chimaeric genes in seed-derived embryos. Plant Cell 1:133–139

Ukai Y, Nishimura S (1987) Regeneration of plants from calli derived from seeds and mature embryos in barley. Jpn J Breed 37:405–411

von Bothmer R (1992) The wild species of *Hordeum*: Relationships and potential use for improvement of cultivated barley. In: Shewry PR (ed) Barley: genetics, biochemistry, molecular biology and biotechnology. Alden Press/CAB International, Oxford, pp 3–18

von Bothmer R, Jacobsen N, Baden C, Jorgensen RB, Linde-Laursen I (1991) An ecogeographical study of the genus *Hordeum*. International Board for Plant Genetic Resources, Rome, pp 127

Walbot V (1988) Preparation of DNA from single rice seedlings. Rice Genet Newslett 5:149–151

Wan Y, Lemaux PG (1994) Generation of large numbers of independently transformed fertile barley plants. Plant Physiol 104:37–48

Weigel RC, Hughes KW (1985) Long term regeneration by somatic embryogenesis in barley (*Hordeum vulgare* L.) tissue cultures derived from apical meristem explants. Plant Cell Tissue Organ Cult 5:151–162

Williams K, Bogacki P, Scott L, Karakousis A, Wallwork H (2001) Mapping a gene for leaf scald resistance in barley line "B87/14" and validation of microsatellite and RFLP markers for marker-assisted selection. Plant Breed 120:301–304

Wu H, McCormac AC, Elliot MC, Chen D (1998) Agrobacterium-mediated stable transformation of cell suspension cultures of barley (*Hordeum vulgare* L.). Plant Cell Tissue Organ Cult 54: 161–171

Xu X, Li B (1994) Fertile transgenic indica rice plants obtained by electroporation of the seed embryo cells. Plant Cell Rep 13:237–242

Yan Q, Zhang X, Shi J, Li J (1990) Green plant regeneration from protoplasts of barley (*Hordeum vulgare* L.). Kexue Tongbao 35:1581–1583

Yao QA, Simion E, William M, Krochko J, Kasha KJ (1997) Biolistic transformation of haploid isolated microspores of barley (*Hordeum vulgare* L.). Genome 40:570–581

Zaghmout OMF (1993) Direct electroporation of plasmid DNA into wheat intact cells of embryogenic callus. Cereal Res Commun 21:301–308

Zimmermann U, Vienken J, Pilwat G (1984) Electrodiffusion of cells. In: Chayen J, Bitensky L (eds) Investigative microtechniques in medicine and biology, vol 1. Dekker, New York, pp 89–167

6 Sorghum Transformation

Z. ZHAO and D. TOMES

6.1 Introduction

Sorghum is essentially grown on marginally fertile land because of its yield stability under adverse conditions with primary distribution in the developing areas of the world, such as Africa and Asia. In the last decade, sorghum (*Sorghum bicolor* L.) was grown on over 100 million acres each year worldwide and current production is nearly 60 million metric tonnes of grain. It ranks as the sixth most planted crop in the world, behind wheat, rice, corn, soybean and barley. Sorghum was first domesticated as food in East Africa several thousand years ago and is now a dietary staple for over half a billion people in more than 30 countries. It has been used both as an ingredient in food and livestock feed. Like other cereal grains, sorghum's reputation as either a food or feed crop suffers because of its low protein quality and low content of essential amino acids such as lysine. Imbalanced amino acid composition of cereals, along with low protein content and inadequate energy intake, contribute to protein-energy malnutrition, especially of children, in many countries in Africa and Asia; this affects not only their physical growth, but also their mental development and immune system.

The low lysine content of sorghum is because the poorly digestible endosperm prolamins are low in lysine. The textbook value of lysine in sorghum (0.22%) is even lower than that of corn (0.26%), whereas tryptophan is higher (0.10 vs. 0.07%) (Cromwell et al. 1998). Hence, unlike corn diets that require supplementation with both lysine and tryptophan, a strategic increase in lysine alone would increase the value of sorghum protein. Improvement of sorghum input traits, such as pest and disease resistance as well as drought tolerance etc., is also needed, especially in the high-yielding areas.

The availability of a routine genetic transformation technology allows breeders to access the genes and technology of the genomics revolution. However, sorghum has been categorized as one of the most recalcitrant monocotyledonous species to manipulate for tissue culture and transformation (Zhu et al. 1998). Sorghum tissue culture and plant regeneration have been successful with certain explants (Masteller and Holden 1970; Ma and Liang 1987; Cai et al. 1987; Cai and Butler 1990; Elkonin et al. 1995; Kaeppler and Pedersen 1997). Although work on transformation of sorghum began about a decade ago, much less success has been achieved compared to other crops. Battraw and Hall (1991) first described genetic transformation of sorghum. They delivered

Molecular Methods of Plant Analysis, Vol. 23
Genetic Transformation of Plants

DNA into protoplasts with electroporation and selected transformed cells, without achieving plant regeneration. Hagio et al. (1991) bombarded sorghum nonregenerable cell suspension and obtained transformed cells. Casas et al. (1993) obtained the first transgenic sorghum plants with microprojectile bombardment of sorghum immature embryos and later on they also obtained transgenic plants with immature inflorescences (Casas et al. 1997). Zhu et al. (1998) and Emani et al. (2002) reported success in sorghum transformation through sorghum callus bombardment. Godwin and Chikwamba (1994) reported inoculation of sorghum meristem tissue with *Agrobacterium*. However, it was ambiguous due to the lack of solid evidence to support stable transformation. Zhao et al. (2000) first demonstrated that *Agrobacterium* was capable of producing transgenic plants of sorghum.

6.2 Sorghum Transformation Process

6.2.1 Plant Materials and Transformation Systems

A choice of plant explants combined with reliable culture systems are prerequisites for tissue culture-based transformation processes in plants, especially in sorghum. A wide range of sorghum explants, either for somatic embryogenesis or for organogenesis, has been tested to establish a dependable regenerable tissue culture system. These explants include immature embryos (Gamborg et al. 1977; Thomas et al. 1977; Dunstan et al. 1978, 1979; Brar et al. 1979; Ma and Liang 1987), mature embryos (Thomas et al. 1977; Cai et al. 1987), immature inflorescences (Brettell et al. 1980; Boyes and Vasil 1984; Cai and Butler 1990; Kaeppler and Pedersen 1997), seedlings (Masteller and Holden 1970; Brar et al. 1979; Davis and Kidd 1980; Smith et al. 1983; Godwin and Chikwamba 1994), leaf fragments (Wernicke and Brettell 1980) and anthers (Rose et al. 1986). However, the success of the genetic transformation achieved in sorghum so far has been established with embryogenic cultures initiated from immature embryos or immature inflorescences. Either immature embryos (Casas et al. 1993; Zhao et al. 2000) or the callus tissue derived from immature embryos (Zhu et al. 1998; Emani et al. 2002) or immature inflorescences (Casas et al. 1997) have been used as DNA delivery targets for transformation. The criteria to evaluate a particular genotype for plant transformation through somatic embryogenesis, especially for cereal crops, usually consist of: (1) high callus initiation frequency, (2) friable and fast-growing callus tissue, (3) maintenance of consistent quality of the callus on medium for an extended time (usually at least for 2–3 months or longer), and (4) regeneration of fertile plants. Friable and fast-growing callus can reduce the length of selection for transformation required prior to plant regeneration. Similarly, robust callus growth and regeneration positively impacts transformation efficiency. However, the last two items are essential for success of plant transformation.

In sorghum, special attention has to be paid to tannin and mucilage production (Cai et al. 1987; Cai and Butler 1990; Kaeppler and Pedersen 1996, 1997). Using the lines that produce none or low pigment and mucilage (Kaeppler and Pedersen 1996, 1997) or manipulation to minimize the production of pigment and mucilage (Cai et al. 1987; Cai and Butler 1990; Elkonin et al. 1995; Zhao et al. 2000) are important for sorghum genetic transformation.

Several DNA delivery methods, such as electroporation, bombardment and *Agrobacterium*, have been successful for sorghum transformation. The selectable marker and level of selection for sorghum appear to be fairly specific. All of the successful examples in sorghum transformation so far, whether using bombardment or *Agrobacterium*, have used the *bar* gene as the selectable marker, driven either by the maize ubiquitin promoter (Christensen et al. 1992; Zhu et al. 1998; Zhao et al. 2000; Emani et al. 2002) or by the CaMV 35S promoter (Gardner et al. 1981; Casas et al. 1993, 1997). Emani et al. (2002) tested the *hph* gene driven by CaMV 35S promoter, however, transgenic plants were not obtained. Early results using the NTPII (kanamycin-resistant) as a selectable marker in sorghum were also unsuccessful (Tomes, unpubl. research). Either bialaphos or phosphinothricin (PPT) was used as a selective agent. Varied concentrations of the selective agents were applied in different stages of the selection process. Usually, a lower concentration of the selective agents was used in the initial selection within the first 2 weeks and then increased selection pressure added to explants for extended time periods. This strategy can be important especially for sorghum transformation because sorghum callus initiation and growth are usually slower than other crops, such as rice (Hiei et al. 1994) and corn (Zhao et al. 2001) A gentle application of the selective agents at the initiation step could reduce the stress and allow new cells to be formed initially and begin the process of regeneration. However, enough selection pressure has to be applied for the major selection process to ensure the propagation of transformed cells only (Casas et al. 1993), otherwise a number of escapes could be regenerated (Emani et al. 2002).

6.2.2 Transformation Via Microprojectile Bombardment

The first transgenic sorghum plants and seed were recovered using microprojectile bombardment as the method of DNA delivery (Casas et al. 1993) in cultivar P89012. Choice of sorghum genotype, application of robust tissue culture techniques, choice of explant tissue and the selectable markers were the most important determinants of successful gene transfer. The genotypes P89012 and TX430 (or hybrids with TX430; Casas et al. 1993; Zhu et al. 1998; Emani et al. 2002) have demonstrated transgenic plants and progeny, perhaps because of the lower levels of phenolic compounds produced from callus from immature embryo and immature inflorescences (see discussion above). Consistent selection pressure using the selectable marker gene *bar* or *pat* with the herbicides bialaphos or phosphinothricin (PPT) has consistently produced transgenic plants. Among transgenic plants and progeny in sorghum, one feature that has been noted is the silencing of genes such as GUS (Casas et al. 1993, 1997; Emani et al. 2002) and rice chitinase (Zhu et al. 1998). The reasons for this propensity for instability of gene expression are somewhat unclear because the number of transgenic events obtained to date is less than ten in any one laboratory using particle gun technology, the presence of complex integration events is often coincident with gene expression variability, and the functional ability of promoters from other cereals has not been adequately characterized in sorghum. Further exploitation of microprojectile bombardment as a genetic engineering tool will depend on improving both the efficiency of transgenic plant recovery and achieving more predictable transgene expression over generations.

6.2.3 *Agrobacterium*-Mediated Transformation

Agrobacterium-mediated cereal transformation is relatively new technology. In the past decade, it has been demonstrated in a number of economically important crops, such as rice (Chan et al. 1992, 1993; Hiei et al. 1994; Rashid et al. 1996; Dong et al. 1996; Lee et al. 1999; Fu et al. 2000; Enriquez-Obregon et al. 2000; Upadhyaya et al. 2000), maize (Ishida et al. 1996; Zhao et al. 2001; Frame et al. 2002), wheat (Cheng et al. 1997), barley (Tingay et al. 1997) and sorghum (Zhao et al. 2000; Zhao et al. 2003). Compared to bombardment, *Agrobacterium*-mediated plant transformation has several advantages. Usually, *Agrobacterium* transformation generates a high proportion of transgenic plants with a low copy number of transgene(s) and a simple insertion pattern (Hiei et al. 1994; Ishida et al. 1996; Cheng et al. 1997; Zhao et al. 2000, 2001). Single copy and simple integration of the transgenes are among the most important considerations in plant transformation. From both a gene expression/silencing and a regulatory point of view, single copy and clear integration of foreign genes into the plant genome are quite critical for product development. In rice (Hiei et al. 1994), maize (Ishida et al. 1996; Zhao et al. 2001) and sorghum (Zhao et al. 2000), transformation efficiencies with *Agrobacterium* harboring the super-binary vector (Komari 1990) are much higher than bombardment. Accessibility of a high throughput transformation system in species such as sorghum will significantly facilitate the increasing needs of supporting plant genomic research as well as speeding up the development of commercial genetically engineered crops. In addition, *Agrobacterium* transformation is a relatively simple procedure compared to bombardment and avoids the preparation of a large aliquot of plasmid DNA or DNA fragments prior to transformation.

A reliable embryogenesis culture system in certain genotype(s) of sorghum (meeting the four criteria mentioned earlier); *Agrobacterium* strains and the super binary vector (Komari 1990); and an effective selectable marker gene are the three essential elements for *Agrobacterium*-mediated sorghum transformation. Like other cereals, *Agrobacterium*-mediated sorghum transformation

Table 6.1. Factors related to the manipulation of *Agrobacterium*-mediated sorghum transformation

Step	Factors	Reference
Infection	Medium components related to *Agrobacterium* infection, i.e., acetocyrigone, glucose etc., *Agrobacterium* concentration, age, physiological status and type, special treatments of explant, i.e., pre-culture, wounding	Sangwan et al. (1991); Bideny et al. (1992); Li et al. (1992); Boase et al. (1998); Cao et al. (1998); Cervera et al. (1998); Vergauwe et al. (1998); Nadolska-Orczyk and Orczyk (2000); Uze et al. (2000); Zhao et al. (2000, 2001)
Co-cultivation	Medium components related to *Agrobacterium* infection and promotion of callus growth, co-cultivation length and temperature	Sangwan et al. (1991); Dillen et al. (1997); Cao et al. (1998); Cervera et al. (1998); Vergauwe et al. (1998); Uze et al. (2000); Zhao et al. (2000, 2001)
Optional resting	Medium components related to promotion of callus growth, i.e., antioxidants, antibiotics, resting length and temperature	Zhao et al. (2000, 2001)
Selection	Medium components related to promotion of callus growth, i.e., antioxidants, antibiotics, type and concentration of selection agent Subculture interval	Sangwan et al. (1991); Cervera et al. (1998); Nadolska-Orczyk and Orczyk (2000); Zhao et al. (2000)
Regeneration	Medium components related to shooting and rooting, selection pressure	Nadolska-Orczyk and Orczyk (2000); Zhao et al. (2000)

consists of five major steps: *Agrobacterium* infection, co-cultivation, resting (optional), selection and plant regeneration. Some of the factors that limit transformation efficiency in sorghum are listed in Table 6.1. The age and the physiological status (sorghum plants grown in the greenhouse or the field) of the embryos as well as the concentration of *Agrobacterium* suspension used for infection impact T-DNA delivery and survival of embryos following infection (Zhao et al. 2000, 2001) had the most dramatic impact on transformation frequency. In general, the early steps in the transformation process, i.e., *Agrobacterium* inoculation, co-cultivation and resting, are very sensitive and the parameters in these steps would affect transformation significantly. However, more important than any one factor in the transformation process, the process of balancing multiple factors against one another is important for developing a workable transformation system in sorghum. *Agrobacterium* infection must deliver enough T-DNA into the target to clearly observe transient expression of a visible marker gene cloned on the binary vector, e.g., GUS (Fig. 6.1), GFP (Stewart 2001) or CRC (Ludwig et al. 1990). The target tissue must be viable and healthy enough to maintain growth following T-DNA delivery.

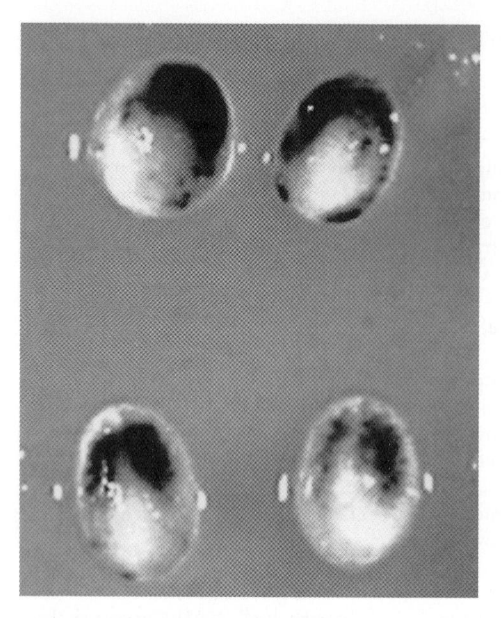

Fig. 6.1. GUS transient expression in sorghum immature embryos following *Agrobacterium* infection and co-cultivation. It can be used to measure T-DNA delivery

6.3 Analysis of Transgenic Plants and the Progeny

Providing genetic, phenotypic and molecular data for the transgenic plants should be considered as part of the transformation research. Solid evidence to prove stable transformation in sorghum is absolutely required, especially at the early stages in working with a new species. Potrykus (1990) listed six requirements to support stable transformation in cereals: (1) serious controls for treatment and analysis, (2) a close correlation between treatment and predicted results, (3) a close correlation between physical (Southern blot, in situ hybridization) and phenotypic (enzyme assays) data, (4) complete Southern analysis containing (a) the predicted signals in high molecular weight DNA in the hybrid fragment between host DNA and foreign gene, and the complete gene, and (b) evidence for the absence of contaminating DNA fragments or the identification of such fragments, (5) data that allow discrimination between false positives and correct treatment in the evaluation of the phenotypic evidence, (6) correlation of the physical and phenotypic evidence with transmission to sexual offspring as well as genetic and molecular analysis of offspring populations. In the past decade, these criteria have been widely accepted and used to measure stable plant transformation.

Although experimental design is required to meet these criteria, the biochemical and genetic analysis can be combined into foreign gene expression and Southern assay in T_0 plants and their sexual progeny.

6.3.1 Molecular Analysis of T_0 Plants

To confirm the integration of foreign gene(s) into the plant genome, molecular assays, such as PCR, in situ hybridization and Southern blots can be used. However, the most widely accepted, reliable and commonly used method is Southern blotting. Usually, three types of enzyme digestion are used to elucidate the insertion of foreign genes into the plant genome and the integration pattern. Uncut genomic transgenic plant DNA hybridizing with the probe that is a unique part or the full sequence of the foreign gene can provide the confirmation of the genomic integration of the foreign gene. However, because the uncut plant genomic DNA stays at the top of the gel, the hybridization bands sometimes were indistinct. Another way to confirm the genomic integration is the so-called Integration Digestion (Fig. 6.2). Digesting the transgenic plant genomic DNA with a restriction enzyme that either has only one cutting site or has no cutting site within the transgene cassette, separating the DNA fragments on a gel and hybridizing these DNA fragments with the probe; either a single band (single insertion locus) or multiple bands (more than one insertion loci) appear clearly. Through this digestion, one can confirm genomic integration of the foreign gene(s) and the number of insertion loci. One can also estimate the copy number in each locus simultaneously if proper copy number controls were used. Three types of controls are needed: nontransformed plant genomic DNA digested with the same enzyme and hybridized with the same probe, plasmid (vector) DNA digested with the same enzyme and hybridized with the same probe and copy number controls (such as 1 copy

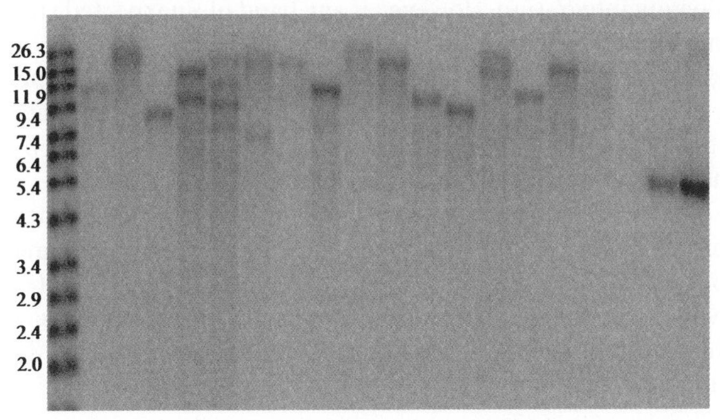

Fig. 6.2. A Southern blot of the genomic DNA isolated from transgenic sorghum T_0 plants produced by *Agrobacterium* transformation and probed with *bar* gene. These genomic DNAs were digested with *Hpa*I that does not have a cutting site within the transgene cassette. *Lane 1* RFLP Ladder System, *lanes 2—17* represent independent events, *lane 18* nontransformed sorghum control and *lanes 19—20* 1 and 5 copies of the plasmid DNA. More than one integration locus occurred in events 5, 6, 7 and 14

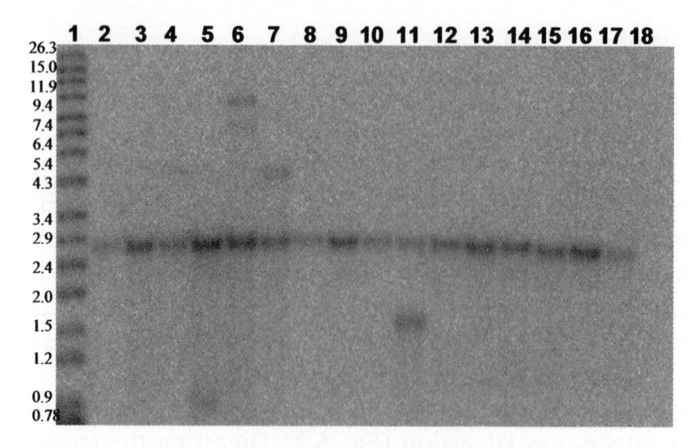

Fig. 6.3. A Southern blot of the genomic DNA isolated from the same transgenic sorghum T_0 plants as in Fig. 6.2 and probed with *bar* gene. These genomic DNAs were digested with *Eco*RV and *Sca*I. *Eco*RV digests the transgene cassette at the 3′ end and *Sca*I digests the transgene cassette near the 5′ end to generate a 2.76-kb fragment. *Lane 1* RFLP Ladder System, *lanes 2–17* represent independent events and *lane 18* nontransformed sorghum control. Rearrangements of the transgene occurred in the events of lane 5, 6, 7 and 11, but not in other events

and 5 copies). In addition, a size ladder marker is needed along with each blot to estimate the size of each band. The third type of digestion is the so-called Plant Transcription Unit (PTU; Fig. 6.3). When digesting the transgenic plant genomic DNA with a restriction enzyme(s) that cuts the DNA at both the 5′ and 3′ ends of the transgene cassette and hybridizing with the probe, a single band with the expected size should be visualized if there is no rearrangement of the transgene following integration. However, if any band of unexpected size is observed, rearrangement(s) within the transgene cassette is likely. For instance, transgene rearrangements occurred in the events of lane 5, 6, 7 and 11, but not in other events in Fig. 6.3. These blots also can be used to estimate the copy number of the insertion. These last two types of Southern blots generate good information to confirm stable transformation.

With *Agrobacterium*-mediated transformation, one of the important issues is the integration of the binary vector backbone sequence into plant genome (Zhao et al. 1998; Buck et al. 2000; Yin and Wang 2000). To elucidate the integration of the vector backbone sequence, Southern analysis or PCR assay against the vector backbone sequence should be performed. This is especially important for development of commercial products.

6.3.2 Foreign Gene Expression in T_0 Plants

Depending on the types of the transgenes used in transformation studies and depending on the purpose of the studies, assays for transgene expression can be qualitative or quantitative. A number of techniques, either at the mRNA level or at the protein level, are used for the assay of foreign gene expression in transgenic plants. These techniques include Northern blotting, RT-PCR Western blotting, ELISA (Rocke and Jones 1997; Crowuther 2001), computer image analysis (Anderson et al. 1994; De Novoa and Coles 1994), histochemical stain (McCabe et al. 1988), visualizing fluorescent protein (Koncz et al. 1987; Chalfie et al. 1994; Harper et al. 1999) or anthocyanin (Ludwing et al. 1990) and herbicide application (Casas et al. 1993).

Recently, transgene silencing has been reported in numerous monocot species (Iyer et al. 2000). It has also been observed in transgenic sorghum plants. Zhu et al. (1998) reported silencing of the integrated chitinase gene in sorghum T_0 plant and progeny. Emani et al. (2002) discussed methylation-based GUS gene silencing in transgenic sorghum plants.

6.3.3 Genetic and Molecular Analysis of the Progeny

Transgene transmission to the progeny through their gametes is important evidence for stable transformation. Both genetic data of transgene segregation and molecular proof of the presence of the transgene(s) in the progeny are needed. Gene expression assay(s) can be used to correlate phenotypic data and physical data.

It has been noted that neither male gametes nor both male and female gametes are fertile in some transgenic plants (Emani et al. 2002). A number of reasons may cause infertile gametes in transgenic plants, such as endogenous gene(s) interrupted by foreign gene insertion, toxicity of the foreign gene product to the gametes, tissue culture conditions or T_0 plant growing conditions. To maximize the possibility of setting seeds in T_0 plants, both self- and outcross pollination (pollination to nontransformed plants) are recommended.

Regardless of the gene silencing issue, to confirm stable transformation, complete Southern analysis in T_0 and T_1 generation is absolutely indispensable.

6.4 Marker-Free Sorghum Transgenic Plants

The elimination of selectable marker genes from transgenic plants, especially in commercial products, has received significant attention over the last decade. In general, the selectable markers are needed in the current technology in most important crops due to the efficiency of transformation. The persistence of the marker genes in the products is undesirable or unacceptable because of the public concern over nonproduct-related genes. Strategies for the efficient elimination of marker genes from transgenic plants have been aggressively pursued over the past decade.

In sorghum, this is more important because of possible cross-pollination between sorghum and weedy species.

6.4.1 Importance of Marker-Free Transgenics in Sorghum

Crop sorghum can spontaneously hybridize with johnsongrass (*Sorghum halepense* L.) under native conditions (Vinall and Getty 1921; Hadley 1953; Baker 1972; Doggett 1988). Johnsongrass is a tetraploid, wind-pollinated perennial weedy plant (Holm et al. 1977). It is self-pollinated and can outcross (Warwick and Black 1983; Kigel and Rubin 1985). Sorghum and johnsongrass occur together in most locations where sorghum is planted and the flowering time of these two species overlaps (Holm et al. 1977). The spontaneous hybridization between crop sorghum, *Sorghum bicolor*, and johnsongrass, *Sorghum halepense*, has been studied extensively (Arriola and Ellstrand 1996).

This situation makes the adoption of genetic engineering for crop sorghum more difficult. The herbicide selectable marker, such as the *bar* gene, is needed in the existing sorghum transformation technology. To avoid the herbicide-resistant gene spreading to weedy species, transgenic sorghum has to be free of herbicide-resistant genes before the wide-scale release of products.

In addition to herbicide-resistant genes, any transformed DNA fragment that is necessary to enhance genetic transformation, but not for plant improvement, should not be included in the final transgenic product. This would include components such as transformation enhancement genes (Lowe et al. 2000), visible markers (e.g., GFP) and antibiotic resistance genes (Walters et al. 1992).

6.4.2 Methods to Eliminate Markers from Transgenic Plants

Several strategies have been developed in plants to eliminate selectable marker genes or other DNA fragments (reviewed by Hohn et al. 2001). These strategies could be divided into five categories: (1) co-transformation, such as co-bombardment or *Agrobacterium* co-transformation systems (Komari et al. 1996). However, usually co-bombardment of two or more plasmid DNAs or DNA fragments produces linked events and the chances of segregating these co-bombarded genes from each other are low. (2) Transposable elements, such as maize Ac/Ds, are used as a vehicle to transpose transgenes from a linked locus to an unlinked locus and eventually these transgenes can be disassociated in the following generation. The feasibility of this strategy has been demonstrated in tomato (Goldsbrough et al. 1993). (3) Site-specific recombination systems (reviewed by Lyznik et al. 2000; Ow 2002), such as the Cre-*lox* system from bacteriophage P1 (Russell et al. 1992) and the FLP-*FRT* system from *Sacchromyces cerevisiae* (Lyznik et al. 1993). have also been used to eliminate marker genes. This strategy can effectively eliminate marker genes in several plant species, such as tobacco (Dale and Ow 1991; Lloyd and Davis 1994; Bar et al. 1996; Gleave et al. 1999), *Arabidopsis* (Russell et al. 1992; Zuo et al. 2001) and maize (Lyznik et al. 1996) as well as tobacco plastids (Corneille et al. 2001; Hajdukiewicz et al. 2001). (4) Special vector design such as the MAT-vector (Ebinuma et al. 2000, 2001; Sugita et al. 1999, 2000) and Double Right Border (DRB) vector (Lu et al. 2001). (5) Use of transgene(s) that are capable of promoting tissue growth and/or plant regeneration (Lowe et al. 2000; Zuo et al. 2002).

In sorghum, most of these strategies have not been tested yet due to the lag in development of efficient technology for genetic transformation.

6.4.3 *Agrobacterium* 2 T-DNA Co-Transformation System

Agrobacterium co-transformation and segregation of transgenes in subsequent generations are strategies developed to eliminate marker genes or other transgenes. This has been evaluated in three ways; (1) transformation with two separate *Agrobacterium* strains, each harboring distinct T-DNAs (Depicker et al. 1985; McKnight et al. 1987; de Block and Debrouwer 1991; de Buck et al. 1998); (2) transformation with a single *Agrobacterium* strain containing two binary vectors (de Framond et al. 1986; Daley et al. 1998) or (3) transformation with a single *Agrobacterium* strain containing two T-DNAs on the same binary vector (Komari et al. 1996). The latter has been successfully adopted in a broad range of plant species including both dicots and monocots, such as tobacco (Komari et al. 1996; McCormac et al. 2001), soybean (Xing et al. 2000), rice (Komari et al. 1996), barley (Matthews et al. 2001) and corn (Miller et al. 2002). This system demonstrated a very reasonable efficiency, ranging from 17 to more than 60%, to generate marker-free transgenic plants in T_1 generation. In addition, it is a simple and straightforward system and does not require additional gene(s) to assist in the elimination of marker genes.

We have made preliminary tests of this system in sorghum (Zhao et al. 2003) through *Agrobacterium*-mediated transformation. A superbinary vector was constructed as described before (Zhao et al. 2000). The T-DNA 1 contained a *bar* gene driven by a maize ubiquitin promoter and pinII terminator and T-DNA 2 contained three copies of *HT-12* gene which encodes a 44-residue protein containing 12 lysine residues (Rao et al. 1994) driven by the maize gamma zein promoter and terminator. Two sorghum lines, P898012 and PHI391, were used as the targets and the same transformation process (Zhao et al. 2000) was used in this study.

With these two sorghum lines, T_0 plants were regenerated in ten stably transformed events. PCR and Southern analysis confirmed that the *bar* gene was present in all of these ten events while the *HT-12* gene was present in five of these ten events. Within these five co-transformed events, *bar* and *HT-12* were segregated in T_1 generation in one of the events and *bar*-free *HT-12* transgenic

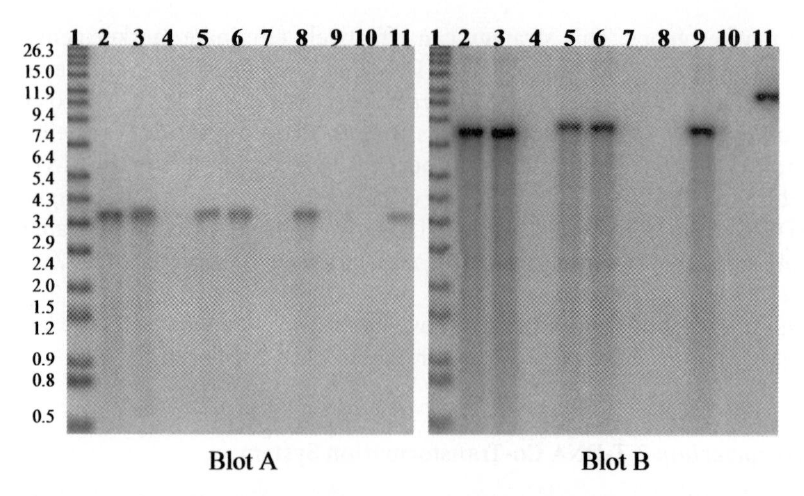

Fig. 6.4. Southern blots of the sorghum events that were co-transformed with *bar* and *HT-12* and showed either segregation or no segregation of these two genes. *Blot A* probed with *bar* gene and *Blot B* probed with *HT-12*. In both blots, *lane 1* RFLP Ladder System, *line 2* a T_0 plant in a non-segregated event showing the presence of both *bar* and *HT-12* genes, *lanes3–4* two T_1 plants of this nonsegregated event, one having both genes (*lane 3*) and one having none (*lane 4*), *lane 5* a T_0 plant in the segregated event showing the presence of both genes, *lanes6–9* four T_1 plants representing each unique genotype, having both genes (*lane 6*), having none (*lane 7*), having *bar* only (*lane 8*), and having *HT-12* only (*lane 9*). *Lane 10* nontransformation control, *lane 11* plasmid DNA control

sorghum plants were recovered in this event (Fig. 6.4). This result is consistent with the data observed in other plant species.

References

Anderson WF, Arthur L, Ozias-Akins P (1994) Quick and efficient measurement of transient GUS expression using image analysis. Plant Mol Biol Rep 12:332–340

Arriola PE, Ellstrand NC (1996) Crop-to-weed gene flow in the genus *Sorghum* (Poaceae): spontaneous interspecific hybridization between johnsongrass, *Sorghum halepense,* and crop sorghum, *S. bicolor.* Am J Bot 83:1153–1160

Baker HG (1972) Migrations of weeds. In: Valentine DH (ed) Taxonomy, phytogeography, and evolution. Academic Press, London, pp 327–347

Bar M, Leshem B, Gilboa N, Gidoni D (1996) Visual characterization of recombination at FLP-*gus*A loci in transgenic tobacco mediated by constitutive expression of the native FLP recombinase. Theor Appl Genet 43:407–413

Battraw M, Hall TC (1991) Stable transformation of *Sorghum bicolor* protoplasts with chimeric neomycin phosphotransferase II and β-glucuronidase genes. Theor Appl Genet 82:161–168

Bidney D, Scelonge C, Martich J, Burrus M, Huffman G (1992) Microprojectile bombardment of plant tissue increases transformation frequency by *Agrobacterium tumefaciens.* Plant Mol Biol 18:301–313

Boase MR, Bradley JM, Borst NK (1998) An improved method for transformation of regal pelargonium (*Pelargonium xdomesticum* Dubonnet) by *Agrobacterium*. Plant Sci 139:59–69

Boyes CJ, Vasil IK (1984) Plant regeneration by somatic embryogenesis from cultured young inflorescences of *Sorghum arundinaceum* (Desv.) Stapf. var. sudanense (susangrass). Plant Sci Lett 35:153–157

Brar DS, Rambold S, Gamborg O, Constabel F (1979) Tissue culture of corn and sorghum. Z Pflanzenphysiol 95:377–388

Brettell RIS, Wernicke W, Thomas E (1980) Embryogenesis from cultured inflorescences of *Sorghum bicolor*. Protoplasma 104:141–148

Buck SD, Wilde CD, Montagou MV, Depicker A (2000) T-DNA vector backbone sequences are frequently integrated into the genome of transgenic plants obtained by *Agrobacterium*-mediated transformation. Mol Breed 6:459–468

Cai T, Butler L (1990) Plant regeneration from embryogenic callus initiated from immature inflorescences of several high-tannin sorghums. Plant Cell Tissue Organ Cult 20:101–110

Cai T, Daly B, Butler L (1987) Callus induction and plant regeneration from shoot portions of mature embryos of high tannin sorghum. Plant Cell Tissue Organ Cult 9:245–252

Cao X, Liu Q, Rowland LJ, Hammerschlag FA (1998) GUS expression in blueberry (*Vaccinium* spp.): factors influencing *Agrobacterium*-mediated gene transfer efficiency. Plant Cell Rep 18: 266–270

Casas AM, Bressan RA, Hasegawa PM (1993) Transgenic sorghum plants via microprojectile bombardment. Proc Natl Acad Sci USA 90:11212–11216

Casas AM, Kononowicz AK, Haan TG, Zhang L, Tomes DT, Bressan RA, Hasegawa PM (1997) Transgenic sorghum plants obtained after microprojectile bombardment of immature inflorescences. In Vitro Cell Dev Biol-Plant 33:92–100

Cervera M, Pina JA, Juarez J, Navarro L, Pena L (1998) *Agrobacterium*-mediated transformation of citrange: factors affecting transformation and regeneration. Plant Cell Rep 18:271–278

Chalfie M, Tu Y, Euskirchen G, Ward WW, Prasher DC (1994) Green fluorescent protein as a marker for gene expression. Science 263:802–805

Chan MT, Lee TM, Chang HH (1992) Transformation of indica rice (*Oryza sativa* L.) mediated by *Agrobacterium*. Plant Cell Physiol 33:577–583

Chan MT, Chang HH, Ho SL, Tong WF, Yu SM (1993) *Agrobacterium*-mediated production of transgenic rice plants expressing a chimeric α-amylase promoter/β-glucuronidase gene. Plant Mol Biol 22:491–506

Cheng M, Fry JE, Pang S, Zhou H, Hironaka CM, Duncan DR, Conner TW, Wan Y (1997) Genetic transformation of wheat mediated by *Agrobacterium tumefaciens*. Plant Physiol 115:971–980

Christensen AH, Sharrock RA, Quail PH (1992) Maize polyubiquitin genes: structure, thermal perturbation of expression and transcript splicing, and promoter activity following transfer to protoplasts by electroporation. Plant Mol Biol 18:675–689

Corneille S, Lutz K, Svab Z, Maliga P (2001) Efficient elimination of selectable marker genes from the plastid genome by the CRE-lox site-specific recombination system. Plant J 27: 171–178

Cromwell GL, Baker DH, Ewan RC, Kornegay ET, Lewis AJ, Pettigrew JE, Steele NC, Thacker PA (1998) Composition of feed ingredients. In: National Research Council Nutrient Requirements of Swine, 10th revised edn. National Academy Press, Washington, DC, pp 124–142

Crowuther JR (2001) The ELISA guidebook. In: Walker JM (ed) Methods in molecular biology, vol 149. Humana Press, Totowa, New Jersey

Dale EC, Ow DW (1991) Gene transfer with subsequent removal of the selection gene from the host genome Proc Natl Acad Sci USA 88:10558–10562

Daley M, Knauf V, Summerfeldt K, Turner J (1998) Co-transformation with one *Agrobacterium tumefaciens* strain containing two binary plasmids as a method for producing marker-free transgenic plants. Plant Cell Rep 17:489–496

Davis ME, Kidd GH (1980) Optimization of sorghum primary callus growth. Z Pflanzenphysiol 98:79–82

De Block M, Debrouwer D (1991) 2 T-DNA's co-transformed into *Brassica napus* by a double *Agrobacterium tumefaciens* infection are mainly integrated at the same locus. Theor Appl Genet 82:257–263

De Buck S, Jacobs A, Van Montague M, Depicker A (1998) *Agrobacterium tumefaciens* transformation and cotransformation frequencies of *Arabidopsis thaliana* root explants and tobacco protoplasts. Mol Plant Microbe Interact 11:449–457

De Framond AJ, Black EW, Chilton WS, Kayes L, Chilton MD (1986) Two unlinked T-DNAs can transform the same tobacco plant cell and segregate in the F1 generation. Mol Gen Genet 202:125–131

De Novoa CO, Coles G (1994) Computer image analysis to quantify and analysis stable transformation identified using the histochemical GUS assay. Plant Mol Biol Rep 12:146–151

Depicker A, Herman L, Jacobs A, Schell J, Van Montague M (1985) Frequencies of simultaneous transformation with different T-DNAs and their relevance to the *Agrobacterium*/plant cell interaction. Mol Gen Genet 201:477–484

Dillen W, Clercq JD, Kapila J, Zambre M, Montagu MV, Angenon G (1997) The effect of temperature *on Agrobacterium tumefaciens*-mediated gene transfer to plants. Plant J 12:1459–1463

Doggett H (1988) Sorghum. Tropical agricultural series, 2nd edn. Longman Scientific, Essex

Dong J, Teng W, Buchholz WG, Hall TC (1996) *Agrobacterium*-mediated transformation of Javanica rice. Mol Breed 2:167–276

Dunstan DI, Short KC, Thomas E (1978) The anatomy of secondary morphogenesis in cultured scutellum tissues of *Sorghum bicolor*. Protoplasma 97:251–260

Dunstan DI, Short KC, Dhaliwal H, Thomas E (1979) Further studies on plantlet production from cultured tissues *of Sorghum bicolor*. Protoplasma 101:355–361

Ebinuma H, Sugita K, Matsunaga E, Endo S, Kasahara T (2000) Selection of marker-free transgenic plants using the oncogenes (IPT, ROL A, B, C) of *Agrobacterium* as selectable markers. In: Jarn SM, Minocha SC (eds) Molecular biology woody plants. Kluwer, Dortrecht, pp 24–46

Ebinuma H, Komamine A (2001) MAT (multi-auto-transformation) vector system. The oncogenes of Agrobacterium as positive markers for regeneration and selection of marker-free transgenic plants. In Vitro Cell Dev Biol 37:103–113

Elkonin LA, Lopushanskaya RF, Pakhomova NV (1995) Initiation and maintenance of friable embryogenic callus of sorghum (*Sorghum bicolor*(L) Moench) by amino acids. Maydica 40: 153–157

Emani C, Sunilkumar G, Rathore KS (2002) Transgene silencing and reactivation in sorghum. Plant Sci 162:181–192

Enriquez-Obregon GA, Prieto-Samsonov DL, de la Riva GA, Perez G, Selman-Housein G, Vazquez-Padron RI (2000) *Agrobacterium*-mediated Japonica rice transformation: a procedure assisted by an antinecrotic treatment. Plant Cell Tissue Organ Cult 59:159–168

Frame BR, Shou H, Chikwamba RK, Zhang Z, Xiang C, Fonger TM, Pegg SEK, Li B, Nettleton DS, Pei D, Wang K (2002) *Agrobacterium tumefaciens*-mediated transformation of maize embryos using a standard binary vector system. Plant Physiol 129:13–22

Fu X, Duc LT, Fontana S, Bong BB, Tinjuangjun P, Sudhakar D, Twyman R, Christou P, Kohli A (2000) Linear transgene constructs lacking vector backbone sequences generate low-copy-number transgenic plants with simple integration patterns. Transgenic Res 9:11–19

Gamborg OL, Shyluk JP, Brar DS, Constabel F (1977) Morphogenesis and plant regeneration from callus of immature embryos of sorghum. Plant Sci Lett 10:67–74

Gardner RC, Howarth AJ, Hahn P, Brown-Luedi M, Shepherd R, Messing J (1981) The complete nucleotide sequence of an infectious clone of cauliflower mosaic virus by M13mp7 shotgun sequencing. Nucleic Acids Res 9:2871–2888

Gleave AP, Mitra DS, Mudge SR, Morris BAM (1999) Selectable marker-free transgenic plants without sexual crossing: transient expression of cre recombinase and use of a conditional lethal dominant gene. Plant Mol Biol 40:223–235

Godwin I, Chikwamba R (1994) Transgenic grain sorghum (*Sorghum bicolor*) plants via *Agrobacterium*. In: Henry RJ, Ronalds JA (eds) Improvement of cereal quality by genetic engineering. Plenum Press, New York, pp 47–53

Goldsbrough A, Lastrella C, Yoder J (1993) Transposition-mediated re-positioning and subsequent elimination of marker genes from transgenic tomato. Bio/Technology 11:1286–1292

Hadley HH (1953) Cytological relationships between *Sorghum vulgare* and *S. halepense*. Agron J 45:139–143

Hagio T, Blowers AD, Earle ED (1991) Stable transformation of sorghum cell cultures after bombardment with DNA-coated microprojectiles. Plant Cell Rep 10:260–264

Hajdukiewicz PT, Gilbertson L, Staub JM (2001) Multiple pathways for Cre/lox-mediated recombination in plastids. Plant J 27:161–170

Harper BK, Mabon SA, Leffel SM, Halfhill MD, Richards HA, Moyer KA, Stewart DN Jr (1999) Green fluorescent protein as a marker for expression of a second gene in transgenic plants. Nat Biotechnol 17:1125–1129

Hiei Y, Ohta S, Komari T, Kumashiro T (1994) Efficient transformation of rice (*Oryza sativa* L.) mediated by *Agrobacterium* and sequence analysis of the boundaries of the T-DNA. Plant J 6:271–282

Hohn B, Levy AA, Puchta H (2001) Elimination of selection markers from transgenic plants. Curr Opin Biotechnol 12:139–143

Holm LG, Plucknett DL, Pancho JF, Herberger JP (1977) The world's worst weeds. University Press of Hawaii, Honolulu, Hawaii

Ishida Y, Saito H, Ohta S, Hiei Y, Komari T, Kumashiro T (1996) High efficiency transformation of maize (*Zea mays* L.) mediated by *Agrobacterium tumefaciens*. Nat Biotechnol 14:745–750

Iyer LM, Kumpatla SP Chandrasekharan MB, Hall TC (2000) Transgene silencing in monocots. Plant Mol Biol 43:323–346

Kaeppler HF, Pedersen JF (1996) Media effects on phenotype of callus cultures initiated from photoperiod-insensitive, elite inbred sorghum lines. Maydica 41:83–89

Kaeppler HF, Pedersen JF (1997) Evaluation of 41 elite and exotic inbred *Sorghum* genotypes for high quality callus production. Plant Cell Tissue Organ Cult 48:71–75

Kigel J, Rubin B (1985) *Sorghum halepense*. In: Halevy AH (ed) CRC handbook of flowering, vol IV. CRC Press, Boca Raton, pp 376–379

Komari T (1990) Transformation of cultured cells of *Chenopodium quinoa* by binary vectors that carry a fragment of DNA from the virulence region of pTiBo542. Plant Cell Rep 9:303–306

Komari T, Hiei Y, Saito Y, Murai N, Kumashiro T (1996) Vectors carrying two separate T-DNAs for co-transformation of higher plants mediated by *Agrobacterium tumefaciens* and segregation of transformants free from selection markers. Plant J 10:165–174

Koncz C, Olsson O, Langridge WHR, Schell J, Szalay AA (1987) Expression and assembly of functional bacterial luciferase in plants. Proc Natl Acad Sci USA 84:131–135

Lee SH, Shon YG, Lee SI, Kim CY, Koo JC, Lim CO, Choi YJ, Han CD, Chung CH, Choe ZR, Cho MJ (1999) Cultivar variability in the *Agrobacterium*-rice cell interaction and plant regeneration. Physiol Plant 107:338–345

Li XQ, Liu CN, Ritchie SW, Peng JY, Gelvin SB, Hodges TK (1992) Factors influencing *Agrobacterium*-mediated transient expression of *gus* A in rice. Plant Mol Biol 20:1037–1048

Lloyd AM, Davis RW (1994) Functional expression of the yeast FLP/FRT site-specific recombination system in *Nicotiana tabacum*. Mol Gen Genet 242:653–657

Lowe K, Abbit A, Glassman K, Gregory C, Hoerster G, Rasco-Gaunt S, Sun X, Lazzen P, Gordon-Kamm W (2000) Use of LEC1 to improve transformation. In Vitro Cell Dev Biol 36:34A, W-15

Lu HJ, Zhou XR, Gong ZX, Upadhyaya NM (2001) Generation of selectable marker-free transgenic rice using double right-border (DRB) binary vectors. Aust J Plant Physiol 28:241–248

Ludwig SR, Bowen B, Beach L, Wessler SR (1990) A regulatory gene as a novel visible marker for maize transformation. Science 247:449–450

Lyznik LA, Nitchell JC, Hirayama L, Hodges TK (1993) Activity of yeast FLP recombinase in maize and rice protoplasts. Nucleic Acids Res 21:969–975

Lyznik LA, Rao KV, Hodges TK (1996) FLP-mediated recombination of *FRT* sites in the maize genome. Nucleic Acids Res 24:3784–3789

Lyznik LA, Peterson D, Zhao ZY, Guan X, Bowen B, Drummond B, Clair GST, Tagliani L, Baszczynski C (2000) Gene transfers mediated by site-specific recombination systems. In: Gelvin SB, Schiperoort RA (eds) Plant molecular biology manual N1, 2nd edn. Kluwer Academic, Dordrecht, pp 1–26

Ma H, Liang GH (1987) Plant regeneration from cultured immature embryos of *Sorghum bicolor* (L.) Moench. Theor Appl Genet 73:389–394

Masteller VJ, Holden DJ (1970) The growth of and organ formation from callus tissue of sorghum. Plant Physiol 45:362–364

Matthews PR, Wang MB, Waterhouse PM, Thornton S, Fieg SJ, Gunler F, Jacksen JV (2001) Marker gene elimination from transgenic barley, using co-transformation with adjacent 'twin T-DNAs' on a standard *Agrobacterium* transformation vector. Mol Breeding 7:195–202

McCabe DE, Swain WF, Martinell BJ, Christou P (1988) Stable transformation of soybean (*Glycine max*) by particle acceleration. Biotechnology 6:923–926

McCormac AC, Fowler MR, Chen DF, Elliott M (2001) Efficient co-transformation of *Nicotiana tabacum* by two independent T-DNAs, the effect of T-DNA size and implications for genetic separation. Transgenic Res 10:143–155

McKnight TD, Lillis MT, Simpson RB (1987) Segregation of genes transferred to one plant cell from two separate *Agrobacterium* strains. Plant Mol Biol 8:439–445

Miller M, Tagliani L, Wang N, Berka B, Bidney D, Zhao ZY (2002) High efficiency transgene segregation in co-transformed maize plants using an *Agrobacterium tumefaciens* 2 T-DNA binary system. Transgenic Res 11:381–396

Nadolska-Orczyk A, Orczyk W (2000) Study of the factors influencing *Agrobacterium*-mediated transformation of pea (*Pisum sativum* L.). Mol Breed 6:185–194

Ow D (2002) Recombinase-directed plant transformation for the post-genomic era. Plant Mol Biol 48:183–200

Potrykus I (1990) Gene transfer to cereals: an assessment. Biotechnology 8:535–543

Rao AG, Hassan M, Hempel JC (1994) Structure-function validation of high lysine analogs of α-hordothionin designed by protein modeling. Protein Eng 7:1485–1493

Rashid H, Yokoi S, Toriyama K, Hinata K (1996) Transgenic plant production mediated by *Agrobacterium* in *Indica* rice. Plant Cell Rep 15:727–730

Rocke DM, Jones G (1997) Optimal design for ELISA and other forms of immunoassay. Technometrics 39:162–170

Rose JB, Dunwell JM, Sunderland N (1986) Anther culture of *Sorghum bicolor*. Plant Cell Tissue Organ Cult 6:15–32

Russell SH, Hoopes JL, Odell JT (1992) Directed excision of a transgene from the plant genome. Mol Gen Genet 234:49–59

Sangwan RS, Bourgeois Y, Sangwan-Norreel BS (1991) Genetic transformation of *Arabidopsis thaliana* zygotic embryos and identification of critical parameters influencing transformation efficiency. Mol Gen Genet 230:475–485

Smith RH, Bhaskaran S, Schertz K (1983) Sorghum plant regeneration from aluminum selection media. Plant Cell Rep 2:129–132

Stewart CN Jr (2001) The utility of green fluorescent protein in transgenic plants. Plant Cell Rep 20:376–382

Sugita K, Matsunaga E, Ebinuma H (1999) Effective selection system for generating marker-free transgenic plants independent of sexual crossing. Plant Cell Rep 18:941–947

Sugita K, Kasahara T, Matsunaga E, Ebinuma H (2000) A transformation vector for the production of marker-free transgenic plants containing a single copy transgene at high frequency. Plant J 22: 461–469

Thomas E, King PJ, Potrykus E (1977) Shoot and embryo-like structure formation from cultured tissues of *Sorghum bicolor*. Naturwissenschaften 64:587

Tingay S, McElroy D, Kalla R, Fieg S, Wang M, Thornton S, Brettell R (1997) *Agrobacterium tumefaciens*-mediated barley transformation. Plant J 11:1369–1376

Upadhyaya NM, Surin B, Ramm K, Gaudron J, Schunmann PHD, Taylor W, Waterhouse PM, Wang MB (2000) *Agrobacterium*-mediated transformation of Australian rice cultivars Jarrah and

Amaroo using modified promoters and selectable markers. Aust J Plant Physiol 27:201–210

Uze M, Potrykus I, Sautter C (2000) Factors influencing T-DNA transfer from *Agrobacterium* to precultured immature wheat embryos (*Triticum aestivum* L.). Cereal Res Commun 28:17–23

Vergauwe A, Geldre EV, Inze D, Montagu MV, Eeckhout EVD (1998) Factors influencing *Agrobacterium tumefaciens*-mediated transformation of *Artemisia annua* L. Plant Cell Rep 18:105–110

Vinall HN, Getty RE (1921) Sudan grass and related plants. USDA Bull 981

Walters D, Vetsch CX, Potts DE, Lundquist RC (1992) Transformation and inheritance of a hygromycin phosphotransferase gene in maize plants. Plant Mol Biol 18:189–200

Warwick SI, Black LD (1983) The biology of Canadian weeds, vol 61. *Sorghum halepense* (L.) Pers. Can J Plant Sci 63:997–1014

Wernicke W, Brettell R (1980) Somatic embryogenesis from *Sorghum bicolor* leaves. Nature 287: 138–139

Xing A, Zhang Z, Sato S, Staswick P, Clemente T (2000) The use of the two T-DNA binary system to derive marker-free transgenic soybeans. In Vitro Cell Dev Biol Plant 36:456–463

Yin Z, Wang GL (2000) Evidence of multiple complex patterns of T-DNA integration into the rice genome. Theor Appl Genet 100:461–470

Zhao ZY, Gu W, Cai T, Tagliani LA, Hondred D, Bond D, Krell S, Rudert ML, Bruce WB, Pierce DA (1998) Molecular analysis of T_0 plants transformed by *Agrobacterium* and comparison of *Agrobacterium*-mediated transformation with bombardment transformation in maize. Maize Genet Coop Newslett 72:34–37

Zhao ZY, Cai T, Tagliani L, Miller M, Wang N, Pang H, Rudert M, Schroeder S, Hondred D, Seltzer J, Pierce D (2000) *Agrobacterium*-mediated sorghum transformation. Plant Mol Biol 44: 789–798

Zhao ZY, Gu W, Cai T, Tagliani L, Hondred D, Bond D, Schroeder S, Rudert M, Pierce D (2001) High throughput genetic transformation mediated by *Agrobacterium tumefaciens* in maize. Mol Breed 8:323–333

Zhao ZY, Glassman K, Sewalt V, Wang N, Miller M, Chang S, Thompson T, Catron S, Wu E, Bidney D, Kedebe Y, Jung R (2003) Nutritionally improved transgenic sorghum. In: Vasil IK (ed) Plant biotechnology 2002 and beyond, Proceedings of 10th IAPTC & B Congress, Kluwer Academic Publishers, Dordrecht, Boston, London, pp 413–416

Zhu H, Muthukrishnan S, Krishnaveni S, Wilde G, Jeoung JM, Liang GH (1998) Biolistic transformation of sorghum using a rice chitinase gene. J Genet Breed 52:243–252

Zuo J, Niu QW, Moller SG, Chua NH (2001) Chemical-regulated, site-specific DNA excision in transgenic plants. Nat Biotechnol 19:157–161

Zuo J, Niu QW, Ikeda Y, Chua NH (2002) Marker-free transformation: increasing transformation frequency by the use of regeneration-promoting genes. Curr Opin Biotechnol 13:173–180

7 Transgenic Sun ower: PEG-Mediated Gene Transfer

P.C. BINSFELD

7.1 Introduction

Sunflower (*Helianthus annuus* L.), as an economical important crop, is relatively young, having been selected and cultivated on a large scale since the latter part of the nineteenth century. Presently, it is the fourth most important oilseed crop in the world. It was grown worldwide on over 21 million ha in 2000, in intermediate, temperate, subtropical and parts of tropical climates. The production of sunflower in the year 2000 was about 26 million tons. Argentina, Russian Federation, Ukraine, European Union, China, the USA and India are the largest sunflower-producing countries, providing more than 75% of the world's sunflower production (FAO 2000).

Sunflower oil has been used widely for human consumption and in industry. It is also fed to livestock because of its high protein content. As a result of rapid population growth in the world and changes in food consumption, the demand for sunflower as an oil-producing crop is likely to rise. Its industrial uses will also expand significantly, since it can be used as a substitute for mineral oil.

7.2 Genetic Variability and Transgenic Breeding

The sunflower has an excellent capacity to adapt to many different environments in the world. Over the past century, breeding and improvement in farming techniques have transformed it into an important oil crop. Conventional breeding techniques, especially hybridization, have dramatically improved the sunflower in quality and quantity, and will play a very important role in sunflower-breeding programs (Friedt 1992; Fick and Miller 1997). Limited land and water resources, expanding populations, environment stresses, and ecological and social demands have caused concern for the long-term production and breeding, given the expected increase in demand. Sexual incompatibility between the crop and wild annual or perennial species limits the genetic pool access for sunflower improvement (Rieseberg et al. 1995; Jan 1997).

Natural (climatic and edaphologic) stresses and diseases are major problems in sunflower production. Diseases are caused by viruses, bacteria,

Molecular Methods of Plant Analysis, Vol. 23
Genetic Transformation of Plants

mycoplasmas, and fungi. Those caused by fungi are the most important ones. Gray stem spot (*Phomopsis helianthi*), downy mildew (*Plasmopara halstedii*), rust (*Puccinia helianthi*), blight (*Alternaria helianthi*) and the *Sclerotinia* infections are some of the most important diseases affecting sunflower cultivation around the world (Gulya et al. 1997). The increasing demand for high quality oil and for industrial purposes requires rapid genetic improvement. The sunflower breeding priorities vary with specific programs, but seed yield, oil content and quality, reducing the seed coat, reduction of chlorogenic acid, high stress tolerance, early ripening, pest and fungal-resistant genotypes should contribute to wider adaptation and an increase in the crop production (Gulya et al. 1997; Degener et al. 1998).

The combination of conventional plant breeding approaches and transgenic breeding methods involving tissue culture techniques, cell biology and recombinant DNA technology lead to better adaptation to the changing demands. Transgenic sunflower production has been restricted for many years by the limitation of available regeneration systems, despite the recent development of efficient procedures for successful protoplast and somatic hybrid regeneration (Krasnyanski and Menzel 1995; Wingender et al. 1996; Henn et al. 1998; Binsfeld et al. 2000). Genetic engineering and interspecific hybridization not only enable the transfer of desirable genes from wild relatives, but also allow the creation and improvement, within a short time, of sunflower varieties containing specific desired traits for a particular metabolite. Transgenic sunflower can also be used as biofactories for production of biopolymers and pharmaceuticals, or molecular farming of antibodies, pharmaceutical proteins, and industrial enzymes (Haq et al. 1995; Cramer et al. 1996; Ponstein et al. 1996). These will add to the value of sunflower production. Finally, the application of biotechnological methods for gene transfer in sunflower provides a powerful tool for tackling fundamental questions in plant biology and plant breeding (Henry 1999).

A gene transfer process is currently a long complex multistage procedure involving three general phases. Phase one involves the selection and application of a delivery system to incorporate the gene or DNA into a viable receptor cell. Phase two involves integration, expression, and selection of the inserted DNA. Finally, phase three involves the recovery and analysis of viable transgenic plants (Hinchee et al. 1994; Birch 1997). Recovery involves tissue culture and regeneration, which is difficult and may limit sunflower transformation.

7.3 Gene Transfer Systems

A gene transfer system involves the use of several technologies which have the capacity to shuttle isolated DNA into a viable host cell. The efficiency of a transfer system is dependent upon the species to be transformed. Although *Agrobacterium* and particle bombardment-mediated gene transfer are the most commonly used systems, they are not the only techniques available. Every system has its advantages and limitations, making continuous development of new systems and improvements upon established systems essential.

7.3.1 PEG-Mediated Gene Transfer

Currently, the PEG (Polyethylene glycol) technique is ranked as the third most common gene transfer system. However, it was the first technique to report the successful integration of foreign genes into a plant cell. This procedure is the most common method for delivering foreign genes or DNA into protoplasts. Protoplasts are ideal for receiving DNA, as cell walls have proven to be a significant barrier in other transformation systems. It is still not completely clear how PEG induces protoplasts to take DNA up, however, some evidence shows a reversible permeabilization of the plasma membrane, which allows the passage of different molecules (Songstad et al. 1995). Molecules ranging in size from small gene sequences or plasmids to large molecules like chromosomes or micronuclei can be transferred into the protoplast in the presence of PEG. This versatility for the transfer of the different molecules makes the PEG system one of the most simple and promising transformation system applicable on a large scale, independent of plant species. Currently, the main limitation of this system is still the low protoplast regeneration ability of most plant species and, therefore, just viable for species whose protoplasts are efficiently regenerated.

Protoplast selection and an optimized cultivation media are the minimal conditions to obtain a fast-growing and highly embryogenic suspension culture for sunflower cell division and plant regeneration from protoplasts (Binsfeld 1999).

PEG-mediated gene transfer in sunflower protoplasts can be achieved by using two different strategies: (1) short DNA molecule uptake (DNA or plasmid) and (2) large DNA molecule uptake (chromosome or micronuclei). PEG-mediated gene transfer appears to be an efficient, reliable, inexpensive and simple system, when it is possible to regenerate plants. A large population of protoplasts can be easily obtained from hypocotyl, mesophyll or a suspension culture, enabling a good chance to obtain independent gene transfer events. The most efficient DNA uptake occurs in the pro-metaphase in actively growing cells (Binsfeld 1999). Regeneration of transgenic plants is possible under in vitro, optimized growth conditions.

7.3.1.1 Short DNA Molecules Uptake

This method consists of a transformation process developed for direct gene transfer, especially for monogenic controlled traits. Although less effective compared to biological vectors, it is attractive due to its simplicity, versatility, low cost and its use in a broad spectrum of species. Basically, it consists of a treatment in which protoplasts are suspended in a DNA-uptake solution containing the desired DNA [DNA-uptake solution: PEG-6000 (8–15%), DMSO (2–6%), 450 mM mannitol, 5 mM spermidine and 80 mM $Ca(NO_3)_2$]. In sunflower protoplasts, the best transient expression or transformation was obtained using 10 μg plasmid DNA per sample of protoplast suspension (1.5×10^6) in 0.5 ml. Add the PEG-uptake solution dropwise to the protoplast suspension and mix gently. Incubate at room temperature for 10 min and add slowly 8 ml sterile wash solution (300 mM KCl, 10 mM $CaCl_2 \times 2H_2O$ and 3 mM MES), mix and centrifuge at 80 ×g for 7 min to remove the DNA-uptake solution. The protoplasts of the pellet are resuspended in growth medium for cell growth and plant regeneration.

The lack of interest in this system is primarily due to the existence of more efficient transformation systems, the low yields of transformants (3–4%) and the inability of many species to be regenerated from protoplasts into viable plants (0.1–0.01%; Kirches et al. 1991). A general recovery of 4–10 transformed calli for 1 million treated protoplasts and 1 regenerate plant for 2 million treated protoplasts.

7.3.1.2 Large DNA Molecule Uptake

The second method consists of PEG-mediated DNA uptake in the form of chromosome segments, micronuclei or microprotoplasts (Fig. 7.1). Transformation using recombinant DNA techniques is very important, but it is limited to monogenic controlled genes and to the few genes that have been identified and cloned. Polygenic determined traits, or traits with unknown molecular background, like disease resistance, stress tolerance or yield, cannot be obtained with this system. Therefore, an alternative transformation system aiming at the transfer of polygenic determined traits, gene families (clusters), for the production of addition lines, for genetic studies or even for recombination or introgression would be of major interest and valuable in most crop breeding programs (Shepard et al. 1983; Ramulu et al. 1995; Binsfeld et al. 2000).

The large DNA molecules (chromosomes, micronuclei, and microprotoplasts) were isolated and prepared as reported by Binsfeld et al. (2000). Basically, this consists of large-molecule uptake called microfusion in which the receptor protoplasts are suspended in a microfusion solution [Microfusion solution: PEG-6000 (4–20%), DMSO (6%), 90 mM mannitol, 25 mM glycine, 5 mM spermidine and 60 mM $CaCl_2 \times 2H_2O$]. The best microfusion results were obtained using micronuclei or microprotoplasts in a relation of 2:1 (micronuclei:protoplasts) at a density of 5.0×10^5 protoplasts in 0.5 ml. The PEG-based microfusion solution was added dropwise to the protoplast suspension and gently mixed. This was incubated at room temperature for 8–10 min. Twelve milliliters sterile wash solution (240 mM KCl, 100 mM $CaCl_2 \times 2H_2O$ and 3 mM MES) was added slowly and incubated for a further 20 min, mixed and cen-

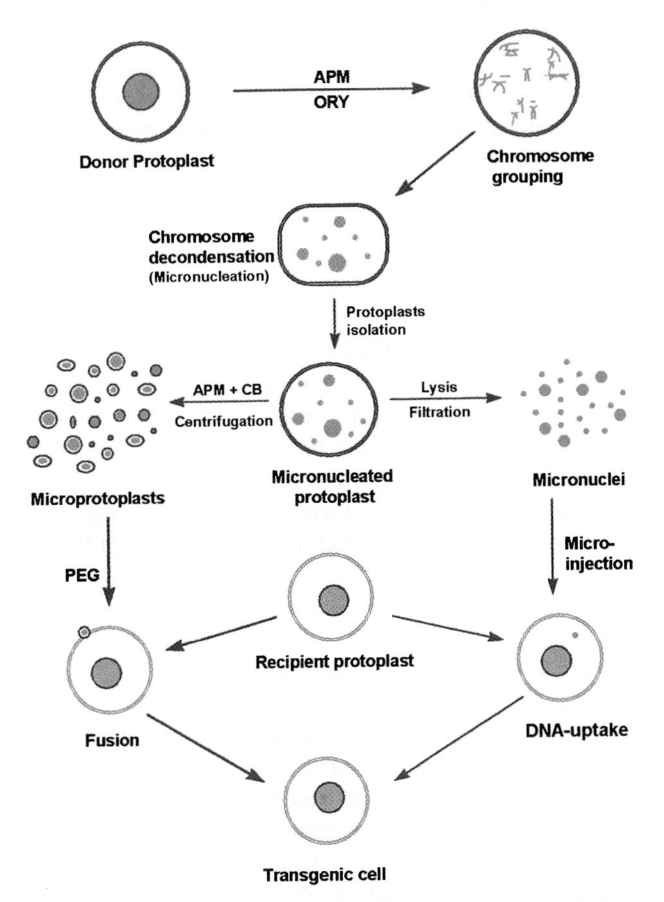

Fig. 7.1. Schematic diagram, showing the basic steps of PEG-mediated large DNA molecule uptake. After micronuclei induction by treatment of donor cells with anti-mitotic toxins (*APM*, *ORY*), chromosomes, micronuclei or microprotoplasts can be isolated from micronucleated protoplasts and transferred to a recipient protoplast

trifuged at 80 ×g for 7 min to remove the microfusion solution. The protoplasts of the pellet were resuspended, counted and cultured in growth medium for cell growth and plant regeneration according to Wingender et al. (1996), Henn et al. (1998) and Binsfeld et al. (1999).

The high interest in this system is primarily due to the possibility of transferring polygenic determined traits and gene families, and of producing additional lines or even foreign gene introgression (Ramulu et al. 1996; Rutgers et al. 1997). Another positive aspect is that a high yield of transformed cells is possible (15–20%), allowing a general recovery of 2–3 transformed plants for 1 million treated protoplasts.

In general, the main limitation of the short or large DNA molecule uptake is related to the low regeneration efficiency of the protoplasts (Binsfeld et al. 2000), therefore, the application of PEG-mediated gene transfer in sunflower is directly related to establishment of an efficient protoplast regeneration protocol.

7.4 Plant Regeneration

Totipotent cells determine the regeneration process. In vitro, callus is the most common tissue used for this purpose. Callus is an amorphous mass of loosely arranged thin-walled parenchyma cells arising from the proliferating cells of the parent tissue. In sunflower, callus can be induced from a variety of tissues or cells, including protoplasts, leaves, stem and roots, making it possible to use these explants for transformation experiments. Totipotent cells with regenerative ability differ between species, but are required for all the different transformation systems.

Most plant transformation protocols involve plant regeneration, which is time-consuming and rate-limiting, as it requires a variety of tissue culture steps and techniques. Thus, tissue culture limitations have prevented some species from being transformed.

Sunflower species are considered as recalcitrant species, i.e., species in which the regeneration process is difficult, but possible using the correct tissues or cells (Binsfeld 1999). Protoplasts were used for the transformation experiments, in which PEG-mediated short or large molecule uptake was tested.

Following the DNA uptake, the protoplasts were cultured according to Wingender et al. (1996), Henn et al. (1998) and Binsfeld (1999). The osmolarity was adjusted to 600 mosmol kg H_2O^{-1} with mannitol. At weekly intervals concomitant to the change of the culture medium, the osmolarity was reduced in steps of 100 mosmol kg H_2O^{-1}. Supplementation with plant growth regulators was done as described by Wingender et al. (1996). Four weeks later, the microcalli that had formed in the agarose droplets were transferred to semi-solid differentiation medium, based on MS salts (Murashige and Skoog 1962) with the following additions: 13 μM glycine, 2.8 mM myo-inositol, 7.5 μM thiamine-HCl, 2 nM nicotinic acid, 1.2 μM pyridoxine-HCl, 5.8 μM silver nitrate, 87.7 mM sucrose, 4 mM MES and 4 g/l phytagel, pH 5.7. Shoot differentiation and plant regeneration occurred in the same culture medium and under growth conditions as described by Wingender et al. (1996) and Henn et al. (1998). Once both shoots and roots had formed, the plants were transferred from the culture medium, potted in soil, and grown in a greenhouse for development and posterior analysis.

7.5 General Analytical Considerations

Once putative transformed tissue is generated, it is essential to test for the newly introduced genes. The proper evaluation and interpretation of the results obtained by PEG-mediated gene transfer requires similar criteria and analytical considerations as for general transgenic plant analysis.

Since by PEG-mediated gene transfer, the desired genetic events (gene transfer) still take place with a low frequency, it is necessary to use and develop efficient analytical methods. Nonetheless, during the regeneration process, many undesirable genetic changes, such as ploidy changes, chromosome rearrangements, silencing and other mutations are also taking place and give rise to callus or plants that are phenotypically indistinguishable from the transgenic plants. These changes may not be rejected by the employed screening methods and thus introduce some uncertainty into the evaluation and interpretation of the events under investigation.

An additional source of error in the interpretation of results, especially by large DNA-molecule uptake, is the production of chimeras (Fromm et al. 1990). The newly introduced genes should be stable, and the transformed tissue nonchimeric. Therefore, it is extremely important to perform analysis in such a way to avoid possible misinterpretations and to provide the means to distinguish true transgenic cells or plants from other mutated cells or plants.

Identification of mutations is extremely important in gene transfer experiments and is complicated as frequently the reversion for most mutations is unknown. In this sense, most cells or plant forms selected by the different screening methods must be subjected to several additional analyses to verify their transgenic origin and to analyze their genetic constitution in more depth. The manipulation and regeneration process can cause several mutations inducing possible misinterpretations of our transgenic plants, in addition to the gene expression in true transgenic plants being unusual.

Several studies have been published in which insufficient attention has been paid to basic considerations of plant analytical methodologies causing equivocal results. Therefore, it is recommended to use a combination of different methods based on molecular, biochemical, cytogenetic, phenotypic and fertility analysis to assure that the results and interpretations are confident as well as viable for transgenic breeding.

7.5.1 Molecular Analysis

The production of transgenic plants requires specific molecular techniques that make characterization at the molecular level possible. These techniques provide precise information about composition and integrity of the inserted DNA; screening; inheritance of the new traits as well as the behavior and expression of the novel protein(s) over time in the transgenic plants.

7.5.1.1 DNA Extraction

DNA extraction for sunflower identification can involve the use of a wide range of protocols (Kirches et al. 1991; Henn et al. 1998; Binsfeld et al. 1999) The nature of the material available for analysis is important as is the type of analysis to be conducted. Dried sunflower material can be used, but young tissue and fresh samples are preferred. Tissues such as those with a high content of oil, phenolics, or starch may produce DNA of a lower quality. The quantity and quality of DNA, as well as its required integrity, will depend upon the analysis. Hybridization-based methods generally require higher amounts of DNA of a much higher quality than PCR-based methods.

7.5.1.2 Southern Hybridization

Southern blot analysis involves the transfer of DNA fragments from an agarose gel to a nylon membrane by capillary transfer and then DNA hybridization with a specific probe to detect the presence of the target DNA fragment (Southern 1975). It offers a sensitive approach for the detection of the target sequence using probes that are labeled. The use of a probe is more sensitive than ethidium bromide detection methods and can reveal a target fragment that was not visible on the original ethidium bromide-stained gel. In addition, when the probe hybridizes it confirms the homology and identity of the fragment.

Transgenic sunflower cells (calli) could be efficiently distinguished by Southern hybridization of a specific probe with the genomic DNA of transformed cells. Ten micrograms of genomic DNA isolated from cells that originated from PEG-mediated short DNA molecule transfer were digested with 4 units *Hin*dIII or *Eco*RI/μg DNA for 3 h at 37°C. The digested DNA was separated by electrophoresis on a 1% w/v agarose gel and blotted onto Hybond N+ nylon membrane according to Sambrook and Russell (2001). A 2.5-kb *Bam*H1–EcoRI fragment containing CaMV promoter and GUS-Nos polyadenilation sequence was excised from pRT103-Gus and used as a probe. The latter was labeled according to the manufacturer's instruction (Amersham, UK) and filters exposed onto X-OMAT films (Kodak) for 24–36 h at –80°C with intensifying screens. High stringency hybridization of the filters was carried out at 65°C overnight followed by three washes (2×, 1×, 0.5× SSC/0.1% v/w SDS) at 65°C for 10 min.

Hybridization of GUS probe to *Hin*dIII-digested genomic DNA demonstrated a consistent uptake and integration of one or more copies of the *gus*A gene into the genome of the transformed cells or plants.

7.5.1.3 Polymerase Chain Reaction

The polymerase chain reaction (PCR) copies DNA in the test tube by using the basic elements of the natural DNA replication process. In a living cell, a highly complex process involving many different proteins replicates the complete genome. PCR uses only the basic components of this complex replication machinery to replicate short fragments of DNA in a test tube. In a simple buffer system, a given region of a large template DNA molecule is copied by DNA polymerase that uses deoxynucleotides as building blocks for the new strands. Sequence-specific primers that bind to the template according to normal base-paring rules define the region of template to be copied. The strands of the template are denatured by heat, which causes the hydrogen bonds between the base pairs of the DNA strands to break. The primers then find their complementary sequences on the templates and DNA polymerase then begins to replicate, producing a new duplex molecule. Usually, it is necessary to carry out between 25–35 cycles to obtain a large amount of a specific DNA strand, which is then used for analysis or further applications (Welsh and McClelland 1990; Caetano-Anolles et al. 1992).

PCR has many advantages over hybridization-based methods for use in plant analysis. These include sensitivity, specificity, speed and ease of application. Using specific primers designed to amplify sequences at distinguishing loci in the genome makes PCR an efficient procedure to identify transgenic sunflower calli or plants.

The efficiency of PEG-mediated short DNA molecule transfer in sunflower was demonstrated by PCR analysis. Genomic DNA of transformed calli using a vector containing the β-glucuronidase (GUS) coding sequence driven by the cauliflower mosaic virus 35S promoter was tested in PCR reactions using specific primers for the *gus*A and CaMV promoter. These primers (P1 = GTGATGGAGCATCAGGGCGGC and P2 = CCTCCTCGGATTCCATTGCCC) amplify a sequence of 600 bp of the *gus*A gene. Reaction mixtures (50 μl) contained 50 μg template DNA, 20 μM dNTPs, 0.2 μM of each primer, 1 unit of Taq polymerase (Repli Pack), 2 mM $MgCl_2$ and 1× reaction buffer. The reactions were run for 35 cycles at 94°C (1 min), 60°C (1 min), and 72°C (2 min).

7.5.1.4 Random Amplified Polymorphic DNA

Random amplified polymorphic DNA (RAPD-PCR) is a relatively rapid PCR-based genomic fingerprinting method. Since RAPD-PCR is based on arbitrary amplification under low restriction conditions, various genomic regions (molecular markers) can be amplified simultaneously in a single PCR amplification (Williams et al. 1990; Caetano-Anolles et al. 1992). These are appropriate to analyze transgenic plants. RAPD-PCR uses a short single primer to initiate DNA synthesis from regions of a template where the primer matches. The initial cycles have to be at low stringency (37–50°C) and the reaction runs for 30–35 cycles.

PEG-mediated large DNA molecule transfer in the form of chromosome segments, micronuclei or microprotoplasts involves various types of genetic changes, i.e., addition lines, aneuploidy, chromosome breakage, deletion, translocation, inversion, introgression as well as punctual molecular changes (Ramulu et al. 1996; Rutgers et al. 1997). In many reports it has been shown that RAPD-PCR generates large numbers of fragments providing several random markers, which are an appropriate tool to analyze such transgenic plants obtained through this DNA uptake system.

Total DNA was extracted from fresh young leaves of the regenerated plants and their respective parent plants and analyzed as described by Binsfeld et al. (1999). For the RAPD PCR reactions, 30 ng of plant DNA was used as template in a final volume of 15 µl reaction buffer, using a modified method of Williams et al. (1990), containing 10 mM Tris-HCl, pH 8.3, 1.5 mM $MgCl_2$, 50 mM KCl, 200 µM dNTPs, 0.4 µM primer and 1 unit of Taq-DNA-polymerase. All primers used were 10-mer random oligonucleotide sequences, series P1, P2, P3, P5, P7 obtained from Pharmacia and I04, I10, I11, B1, B5, B12, B18, D1, D3, D13, D20, P160–2, P160–6, P170–4, P170–10 from Roth Random Primer (Roth, Germany). The reaction samples were overlaid with 10 µl mineral oil (Sigma-Aldrich, Germany) and amplification was performed in 0.5-ml reaction tubes by an AutogeneII thermocycler (Grant, GB). The reactions were run for one start cycle at 94°C (2 min), followed by 35 cycles at 94°C (12 s), 36°C (48 s), 72°C (1.5 min) and a final extension cycle at 72°C (10 min). The PCR products were separated by electrophoresis on 1.5% agarose gels containing ethidium bromide (2.5 µM) in 1× Tris-borate-EDTA (TBE) buffer. Gels were photographed under UV light (302 nm).

7.5.2 Biochemical Analysis

Biochemical analysis involves the determination of gene expression in transgenic cells or plants as a result of stable integration and functional enzymatic activity in transgenic cells or plants obtained by PEG-mediated gene transfer. Biochemical methods have proved to be very informative for transgenic plants, especially in those plants in which DNA integration was demonstrated by molecular analysis.

7.5.2.1 Multiple Molecular Forms of Enzymes

Isoenzymes are variants of an enzyme occurring within a single species. These enzymes have similar general catalytic activity, but differ in physical proprieties (e.g., size, stability, isoelectric point, optimum pH) and are easily separated by gel electrophoresis. Isoenzymes represent a genetically distinct protein encoded at different loci or multiple alleles at a single genetic locus. This makes isoenzymes an interesting tool for genetic variability analysis (Scandalios 1969). Isoenzymes can work as genetic markers and provide information about the interaction and expression of particular genes present in particular chromosomes or micronuclei.

Isoenzyme analysis can be a simple, easy and informative method for transgenic plant detection, especially those plants obtained by PEG-mediated large DNA molecule transfer (i.e., chromosomes or micronuclei). Because such transgenic plants have an additive character, and gene expression of the donor genotype is expected, we analyzed putative transgenic tissue for the following isoenzymes: esterase, acid phosphatase, peroxidase and malate dehydrogenase. Proteins were isolated in a buffer solution containing 0.05 M TRIS, 0.01 M β-mercaptoethanol, and 12% glycerol (pH 7.5). Electrophoretical separation of the proteins was carried out on 10% polyacrylamide gels. Isoenzyme patterns of the putative transgenic tissues showed a low elucidative indication of the nature and level of the specific response that can be attributed to the transferred chromosomes. These results may be interpreted as gene silencing or an indication that the specific genes for the analyzed enzymes are located on different chromosomes to the transferred ones.

7.5.2.2 Enzymatic Assay

The pRT103-Gus vector containing the gene (*gus*A) that expresses β-glucuronidase (GUS) was transferred to sunflower protoplasts by PEG-mediated small DNA molecule uptake technique. The enzymatic assays used for GUS activity determination in transformed tissues are a very convenient tool and have been widely used for analysis of transient or stable gene expression in transgenic plants (Brasileiro and Carneiro 1998). They are also used for transformation protocols optimization, as well as for monitoring gene stability over time and to evaluate structure–function relations of gene-promoter constructs in many plant species.

In sunflower, the β-glucuronidase gene was used as a marker gene for analysis of gene expression in transformed cells. GUS activity can be measured using fluorometric analysis of very small amounts of transformed tissue. The enzyme β-glucuronidase is very stable, and tissue extracts continue to show high levels of GUS activity after prolonged storage at –80°C, but we should avoid storage of the extracts at –20°C as it seems to inactivate the enzyme. Expression of *gus*A genes might be influenced by the physiological differences in transgenic tissues; therefore, care must be taken to choose uniform plant material for the GUS activity in transgenic cells for the GUS assay.

The activity of β-glucuronidase can be determined by colorimetric or fluorometric assays, these procedures in the most plant transformation laboratories are used as a standard analysis. These are rapid, easy, inexpensive and accessible methods.

Colorimetric Determination. Histochemical localization of β-glucuronidase (indicated by dark blue color) was done as described by Jefferson et al. (1987). In this reaction, 5-bromo-4-chloro-3-indolyl-β-D-glucuronide (X-Gluc) is cleaved to yield the blue precipitate, dichloro-dibromoindigo. After transformation, the protoplasts are cultivated for 6–8 weeks producing a microcallus originated from one cell. This callus is then divided into two parts; one part is then used for GUS activity measurement. The callus was soaked in a solution of 2 mM X-Gluc in 50 mM phosphate buffer, pH 7 and incubated at room temperature overnight. After incubation, the presence or absence of blue color is scored. Although GUS activity can also be detected with a more sensitive fluorometric assay, for most transformation experiments and for monitoring purposes, the colorimetric assay is sufficient and less labor-intensive.

Fluorometric Determination. The β-glucuronidase enzyme kinetics were determined using the substrate 4-methylumbelliferyl-β-D-glucuronide (MUG) that cleaves the β-glucuronidase to yield the fluorescent product 7-hydroxymethylcuomarin (MU). The *gusA* gene expression was fluorometrically quantified as described by Jefferson et al. (1987). The protein extracts were added to the GUS assay buffer (5 mM MUG in extraction buffer) and incubated at 37°C for 30 min. After the incubation period, the addition of 0.2 M Na_2CO_3 stopped the reaction, and the fluorescence emission was measured on a fluorometer, JY3D (Jobin Yvon Instruments, Germany), at 365 nm excitation and 455 nm emission. The GUS activity for samples was calculated from a calibration curve with MU as the standard.

7.5.3 Cytogenetic Analysis

Cytogenetic studies are done at the nuclei and chromosomal level by using techniques that yield specific information about the transgenic plants obtained through PEG-mediated large DNA molecule transfer in the form of chromosomes segments, micronuclei or microprotoplasts. This transformation technique favors several types of genetic changes, like addition lines, aneuploidy, chromosome breakage, deletion, translocation, inversion, introgression as well as punctual molecular changes (Sybenga 1992; Ramulu et al. 1996). Since it is possible to obtain a large diversity in nuclear constitution, knowledge about behavior, stability, roles and effects of the introduced large molecules in the putative transgenic plants is essential when aiming at practical breeding purposes.

7.5.3.1 Flow Cytometric Analysis

Flow cytometry involves the analysis of the fluorescence and light scatter properties of single particles of nucleic acid during their passage within a narrow, precisely defined liquid stream. This technique allows fast and accurate quantitative and qualitative analysis of nuclear DNA including chromosomes, micronuclei and nuclei, by measuring the fluorescence of fluorochrome that is specifically bound to the DNA (e.g., propidium iodide, DAPI). The technique consists of the intact nuclei isolation from a large cell population. The nuclei are separated from cell debris by centrifugation or filtration and stained with a fluorochrome. The stained suspension is then injected into the flow cytometer. Nuclei are taken into a tube via a current of water or buffer solution and passed, one by one, across a tight beam of light filtered to the absorption wavelength of the fluorochrome. The fluorochromes bound to the DNA fluoresce, and the emitted wavelength is read by a series of photomultipliers. There is a linear correlation between the amount of DNA in the nuclei and the intensity of the emitted wavelength (Dolezel et al. 1994; Galbraith 1994).

Flow cytometry can be used to measure small DNA molecules, down to 20 kb, below the threshold of separation by normal electrophoresis. However, the ability to accurately quantify large (from 20 kb to chromosomal size) DNA molecules transformed flow cytometry into a very important tool for analysis of large genomic features. Genome size measurements (including the specific AT and GC base-pair content), cell cycle analysis, flow karyotyping (by measuring the DNA content of chromosomes), chromosome sorting and production of chromosome-enriched DNA libraries, ploidy and aneuploidy changes are common examples for transgenic plants analysis obtained through intra- and interspecific large molecule genetic manipulation (Galbraith 1994; Binsfeld et al. 2000).

With recent technical improvements in modern flow cytometers, it is now easy for a researcher to become confident with the technique. A small amount of living material can be collected at any growth stage of the plant. User bias is virtually eliminated. Samples can be run in bulk, greatly increasing the amount of material that can be examined, and internal standards can be included for use in correcting shifting effects and determining DNA content. Furthermore, tens of thousands of cells can be counted in a single run, producing data with low statistical error (Marie and Brown 1993; Dolezel et al. 1994)

Flow cytometric analysis was performed for genome size, ploidy level and relative DNA content determination of interphase nuclei of the parental and regenerated transgenic sunflower plants, based on the method described by Nagl and Treviranus (1995) and Ayele et al. (1996). The isolation of the nuclei for DNA determination was performed with small samples (10 µl of freshly isolated protoplasts), using 1 ml of ice-cold commercial chopping solution (Solution A of Partec GmbH kit). Five minutes later, the suspension with the isolated nuclei was mixed and stained with 1 ml of 4,6-diamino-2-phenylindole (DAPI, Solution B of Partec GmbH Kit) for 5 min. As a standard, nuclei of *Petunia hybrida* cv. F_1 hybrid "Hit parade blau" were used (2C = 2.85 pg, Marie and Brown 1993; Nagl and Treviranus 1995). The samples were measured on a Partec CA-III flow cytometer (Partec, Münster, Germany), equipped with an HBO-100 mercury high pressure lamp, with the UG1 excitation filter, TK420, TK560 dicroic mirrors and CG435 long-pass filter. The software DPAC (Data pool application for cytometry – Partec) was used for the calculation of CV-values and evaluation of the diagrams of relative DNA content.

7.5.3.2 Mitotic and Meiotic Cell Analysis

Metaphase chromosome analysis of mitotic and meiotic cells such as chromosome constitution, chromosome banding, in situ hybridization, and fluorescent tagging methods help us to make important inferences about structure, rearrangement, recombination and behavior of homologous chromosomes in transgenic cells or plants as a result of genetic manipulation and chromosome engineering. Because of the additive nature of PEG-mediated large DNA molecule (chromosome or micronuclei) transfer, it can be expected to produce transgenic plants with chromosomal anomalies as well as chromosome elimination or reduced fertility. In such plants, disturbances in mitosis and meiosis frequently produce several cytogenetic anomalies. These anomalies, e.g., anaphase bridges, laggard chromosomes and univalent or multivalent chromosome pairing at diakinesis, partial chromosome disjunction, individual condensation of the separated chromosomes, finally led to the formation of micronuclei in the tetrads resulting in small and sterile pollen grains. Despite the wide spectrum of chromosomal abnormalities, normal bivalent chromosome pairing is the rule, so that no systematic trend for chromosome elimination can be expected (Jacobsen et al. 1995; Jan 1997).

Chromosome analysis of mitotically active root-tip cells from the regenerated transgenic plants and their parents were collected and pre-treated in an aqueous solution (2 ml) of 2.5 mM 8-hydroxyquinoline for 3 h at 4°C, fixed in (2 ml) 3:1 (v/v) ethanol-acetic acid for 36 h at 4°C, and stored in (2 ml) 70% ethanol at 4°C. Before maceration, the root tips were incubated in 100 µl of enzyme mixture (4% cellulase, Onozuka R-10, Serva, 1% pectolyase Y-23, Seishim Pharmaceutical, 75 mM KCl, pH 4.0; Kakeda et al. 1991) for 30 min at 37°C. For squashing, 45% acetic acid and for staining, DAPI (10 mM) and carmine acetic acid (50%) is recommended.

For meiotic analysis flower buds of transgenic plants were collected and fixed in 15 ml of 3/1 (v/v) ethanol/glacial acetic acid for 36 h at room temperature and stored at 4°C in 15 ml of 70% ethanol. Anthers were digested in a 100-µl enzyme mixture [4% cellulase, Onozuka R-10, Serva and 1% pectolyase (Y-23, Seishim Pharmaceutical) in 75 mM KCl, pH 4.0 (Kakeda et al. 1991)], for 20 min at 37°C and then squashed in 45% acetic acid. The samples are stained with a drop (5 µl) of 10 µM DAPI for fluorescence microscope analysis or with a drop of 50% carmine acetic acid for light microscopy. The investigation of all meiotic phases allows us to identify anomalies during chromosome migration, such as laggard chromosomes, anaphase bridges and their segregation at diakinesis as well as univalent, bivalent and multivalent chromosome pairing. Photographs were taken using a features computer-assisted cooled CCD camera (Photometrics).

7.5.3.3 In Situ Hybridization

In situ hybridization (ISH) is a technique widely used to localize and determine the distribution of particular gene sequences of chromosomes or chromosome regions. ISH refers to the hybridization of nucleic acid probes to cytological preparations of intact chromosomes and chromatin. In plant genomics, the development of nucleic acid detection by fluorescence in situ hybridization (FISH) is becoming a powerful technique for transgenic plant detection, determination of gene localization, number of gene copies in the genome as well as gene or chromosome mapping (Garriga-Calderé et al. 1999).

Genome composition of transgenic plants was analyzed by genomic in situ hybridization following methods described by Kuipers et al. (1997). Chromosome spreads were produced from protoplast suspensions from root tips of young plants, then fixed in 4% formaldehyde for 10 min and denatured using standard techniques. Probes were directly labeled with the Cy3-avidin-streptavidin system by Nick translation and resuspended in hybridization solution with 40-fold excess of unlabelled, sheared *H. annuus* competitor DNA. The probe was pre-annealed for 30 min at 37°C prior to hybridization. The slides were transferred to a moist chamber for hybridization at 37°C overnight. Slides were counterstained with DAPI (4,6-diamino-2-phenylindole) and examined using a fluorescence microscope.

7.5.4 Morphological Analysis

Evaluation of a transgenic effect is a key component of transgenic plant analysis and transgenic breeding. Evaluation strategies and procedures can vary widely and depend on the genes and traits considered. It can be carried out at different levels, but at the expression level (ultimate trait and product) is the most important. Therefore, transgenic plants should be routinely analyzed under controlled or field growth conditions, observing alteration on gene expression corresponding to the expected traits or morphological differences in relation to the parent plants. In our experiments, the regenerated plants were analyzed for several morphological traits, like growth habit, stem, leaves, inflorescence, flowers, and seeds as well as the parental plants (cultivar Florum-328) were analyzed under field conditions.

Pollen fertility and viability was assessed by differential staining of viable and nonviable pollen grains as described by Alexander (1980) and by a pollen germination technique based on a mixture of 1.5% sucrose, 0.1% agar, and 0.005% boric acid. Pollen grains from five anthers, collected from different flowers of the head, were suspended in a drop (20 μl) of staining solution and distributed on four slides. After 30 min the viable and nonviable pollen grains can be determined by differential staining.

Disease resistance test as well as mycelial growth of *Sclerotinia sclerotiorum* were conducted as previously described by Henn et al. (1997). Stem inoculation was performed after 5–6 days of mycelial growth, plugs of 0.5 cm in diameter of mycelia were cut, attached and fastened with Parafilm (Sigma) between the fourth and fifth stem segment of transgenic plants growing in greenhouse. After 6 days, disease symptoms were scored and the Disease Severity Index (DSI) was calculated for the transgenic and the respective control plants.

7.6 Conclusions and Future Perspectives

As the methods discussed here and those under development are refined, it is clear that steady increases in transformation efficiency will occur. The PEG-mediated transformation system for sunflower has shown to be effective for transgenic plant production. However, the transformation efficiency obtained with both large or short molecule transfer system was low, in general 1–2 transformed plants for 1 million treated protoplasts. This suggests that a higher number of protoplasts need to be used initially to ensure a significant number of transformed plants. There might be some possibilities of improving the efficiency of the transformation by refining the delivery system and the regeneration conditions. In addition, accurate selection tools and analytical diagnostics are crucial for a correct interpretation of the results as well for success in transgenic breeding and transgenic plant monitoring.

It appears that transgene technology will play a crucial role in sunflower breeding in the future. Key issues for the success of transgenic breeding include the knowledge and availability of efficient transformation technologies that make genetic engineering routine; moving genes of interest across different biological species, with the possibility to express them in desired phenotypes and acceptable genetic behavior and compatibility of the introduced transgenes with endogenous genes in host genomes. With the rapid advancement of transgene breeding, molecular markers and genomics research, a universal gene pool will be available for sunflower breeding. In parallel, significant progress has been made in the development of analytical technology to identify and monitor the transgenes in the host genomes. Compared to the progress made in the areas of gene discovery and delivery, our knowledge in the analytical area must be continuously improved, allowing precise inferences and understanding of the biochemical complexity, gene interactions and the influence of several factors on transgenic expression in plants. Lack of knowledge in analytical technologies will no doubt present a significant barrier to effective use of transgene technology in breeding. Fortunately, both academic and practical researchers are increasingly realizing the importance of these issues and some significant methodological progress has been made as discussed in this chapter. It can be certain that our understanding of transformation processes and transgenic breeding will continue to be enriched as more transgenic plants are produced and thoroughly characterized.

References

Alexander MP (1980) A versatile stain for pollen from fungi, yeast and bacteria. Stain Technol 55:13–18

Ayele M, Dolezel J, Van Duren M, Brunner H, Zapata-Arias FJ (1996) Flow cytometric analysis of nuclear genome of the Ethiopian cereal Tef [*Eragrostis tef* (Zucc.) Troter]. Genetica 98:211–215

Binsfeld PC (1999) Interspecific transgenic plants in the genus *Helianthus*. Cuvillier, Göttingen, 166 pp

Binsfeld PC, Wingender R, Schnabl H (1999) Direct embryogenesis in the genus *Helianthus* and RAPD analysis of obtained clones. J Appl Bot 73:63–68

Binsfeld PC, Wingender R, Schnabl H (2000) Characterization and molecular analysis of transgenic plants obtained by microprotoplast fusion in sunflower. Theor Appl Genet 101:1250–1258

Birch RG (1997) Plant transformation: problems and strategies for practical application. Annu Rev Plant Physiol Plant Mol Biol 48:297–326

Brasileiro ACM, Carneiro VTC (1998) Manual de transformação genética de plantas. Embrapa-SIP/Embrapa-Cenargen, Brasília, 309 pp

Caetano-Anolles G, Bassam BJ, Gresshoff PM (1992) DNA amplification fingerprinting using very short arbitrary oligonucleotide primers. Bio/Technology 9:553–557

Cramer CL, Weissenborn DL, Oishi KK, Grabau EA, Bennett S, Ponce E, Grabowski GA, Radin DN (1996) Bioproduction of human enzymes in transgenic tobacco. Ann NY Acad Sci 792:62–71

Degener J, Melchinger AE, Gumber RK, Hahn V (1998) Breeding for *Sclerotinia* resistance in sunflower: a modified screening test and assessment of genetic variation in current germplasm. Plant Breed 117:367–372

Dolezel J, Lucretti S, Schubert I (1994) Plant chromosome analysis and sorting by flow cytometry. Crit Rev Plant Sci 13:275–309

FAO (2002) http://apps1.fao.org (authorization necessary)

Fick GN, Miller JF (1997) Sunflower breeding. In: Schneiter AA (ed) Sunflower technology and production. Agronomy, vol 35. Madison, Wisconsin, pp 395–439

Friedt W (1992) Present state and future prospects of biotechnology in sunflower breeding. Field Crop Res 30:425–442

Fromm ME, Morrish F, Armstrong C, Williams R, Thomas J, Klein TM (1990) Inheritance and expression of chimeric genes in the progeny of transgenic maize plants. Biotechnology 8: 833–839

Galbraith DW (1994) Flow cytometry and sorting of plant protoplasts and cells. Methods Cell Biol 42:539–561

Garriga-Calderé F, Huigen DJ, Jacobsen E, Ramanna MS (1999) Prospects for introgression tomato chromosomes into the potato genome: an assessment through GISH analysis. Genome 42: 282–288

Gulya T, Rashid KY, Masirevic SM (1997) Sunflower diseases. In: Schneiter AA (ed) Sunflower technology and production. Agronomy, vol 35. Madison, Wisconsin, pp 263–379

Haq TA, Mason HS, Clements JD, Arntzen CJ (1995) Oral immunization with a recombinant bacterial antigen produced in transgenic plants. Science 268:714–716

Henn H-J, Wingender R, Schnabl H (1997) Wildtype sunflower clones: source for resistance against *Sclerotinia sclerotiorum* (Lib.) de Bary stem infection. J Appl Bot 71:5–9

Henn H-J, Wingender R, Schnabl H (1998) Regeneration of fertile interspecific hybrids from cell fusion between *Helianthus annuus* L. and wild *Helianthus* species. Plant Cell Rep 18:220–224

Henry RJ (1999) Molecular techniques for the identification of plants. In: Chopra VL, Malik VS, Bhat SR (eds) Applied plant biotechnology. Science Publishers, New Hampshire, pp 269–283

Hinchee MAW, Corbin DR, Armstrong CL, Fry JE, Sato SS, DeBoer DL, Petersen WL, Armstrong TA, Conner-Ward DV, Layton JG, Horsch RB (1994) Plant transformation. Plant Cell Tissue Cult 38:231–270

Jacobsen E, De Jong JH, Kamstra SA, Van den Berg PMMM, Ramanna MS (1995) Genomic in situ hybridization (GISH) and RFLP analysis for the identification of alien chromosomes in the backcross progeny of potato (+) tomato fusion hybrids. Heredity 74:250–257

Jan CC (1997) Cytology and interspecific hybridization. In: Schneiter AA (ed) Sunflower technology and production. Agronomy, vol 35. Madison, Wisconsin, pp 497–558

Jefferson RA, Kavanagh TA, Bevan MW (1987) GUS fusion: β-glucoronidase as a sensitive and versatile gene fusion marker in high plants. EMBO J 6:3901–3907

Kakeda K, Fukui K, Yamagata H (1991) Heterochromatic differentiation in barley chromosomes revealed by C- and N-banding. Ther Appl Genet 81:144–150

Kirches E, Frey N, Schnabl H (1991) Transient expression in sunflower mesophyll protoplast. Bot Acta 104:212–216

Krasnyanski S, Menczel L (1995) Production of fertile somatic hybrid plants of sunflower and *Helianthus giganteus* L. by protoplast fusion. Plant Cell Rep 11:7–10

Kuipers AGJ, van Os DPM, de Jong JH, Ramanna MS (1997) Molecular cytogenetics of *Alstromeria*: identification of parental genomes in interspecific hybrids and characterization of repetitive DNA families in constitutive heterochromatin. Chromos Res 5:31–39

Marie D, Brown SC (1993) A cytometric exercise in plant DNA histograms, with 2C values of 70 species. Biol Cell 78:41–51

Murashige T, Skoog F (1962) A revised medium for rapid growth and biossays with tobacco tissue culture. Physiol Plant 15:473–497

Nagl W, Treviranus A (1995) A flow cytometric analysis of the nuclear 2C DNA content in 17 *Phaseolus* species (53 genotypes). Bot Acta 108:403–406

Ponstein AS, Verwoerd TC, Pen J (1996) Production of enzymes for industrial use. Ann NY Acad Sci 792:91–98

Ramulu KS, Dijkhuis P, Rutgers E, Blass J, Verbeek WHJ, Verhoeven HA, Colijn-Hooymans CM (1995) Microprotoplast fusion technique: a new tool for gene transfer between sexually-incongruent plant species. Euphytica 85:255–268

Ramulu KS, Dijkhuis P, Rutgers E, Blass J, Krens FA, Verbeek WHJ, Colijn-Hooymans CM, Verhoeven HA (1996) Intergeneric transfer of a partial genome and direct production of monosomic addition plants by microprotoplast fusion. Theor Appl Genet 92:316–332

Rieseberg LH, Randal Linder C, Seiler GJ (1995) Chromosomal and genetic barriers to introgression in *Helianthus*. Genetics 141:1163–1171

Rutgers E, Ramulu KS, Dijkhuis P, Blaas J, Krens FA, Verhoeven HA (1997) Identification and molecular analysis of transgenic potato chromosomes transferred to tomato through microprotoplast fusion. Theor Appl Genet 94:1053–1059

Sambrook J, Russell DW (2001) Molecular cloning: a laboratory manual, 3rd edn. Cold Spring Harbor Laboratory Press, Cold Spring Harbor, New York

Scandalios JG (1969) Genetic control of multiple molecular forms of isoenzymes in plants – a review. Biochem Genet 3:37–79

Shepard JF, Bidney D, Barby T, Kemble R (1983) Genetic transfer in plant through interspecific protoplast fusion. Science 219:283–688

Songstad DD, Somers DA, Griesbach RJ (1995) Advances in alternative DNA delivery techniques. Plant Cell Tissue Organ Cult 40:1–15

Southern EM (1975) Detection of specific sequences among DNA fragments separated by gel electrophoresis. J Mol Biol 98:503–517

Sybenga J (1992) Cytogenetics in plant breeding. Monographs on theoretical and applied genetics 17. Springer, Berlin Heidelberg New York, 469 pp

Welsh J, McClelland M (1990) Fingerprinting genomes using PCR with arbitrary primers. Nucleic Acids Res 18:7213–7218

Williams JGK, Kubelik AR, Livak J, Rafalski JA, Tingey SV (1990) DNA polymorphisms amplified by arbitrary primers are useful as genetic markers. Nucleic Acids Res 18:6531–6535

Wingender R, Henn H-J, Barth S, Voeste D, Machlab H, Schnabl H (1996) Regeneration protocol for sunflower (*Helianthus annuus* L.) protoplasts. Plant Cell Rep 15:742–745

8 Transformation of Norway Spruce (*Picea abies*) by Particle Bombardment

D.H. Clapham, H. Häggman, M. Elfstrand, T. Aronen, and S. von Arnold

8.1 Introduction

Norway spruce (*Picea abies*) and Scots pine are the most important European economic forest tree species. Of the 112 million ha of forest in northern and western Europe, Sweden and Finland contribute 44 million ha, of which 40–45% consist of Norway spruce and a similar area consists of Scots pine (FAO 1999). The demand for forest products is expected to increase over the next 50 years, by about 1.7% per year (Fenning and Gershenzon 2002). This increase in demand will have to be met by increased production, for the most part on the same or smaller area. Here, forest tree breeding plays an important role. In Scandinavia, the long rotation cycles of 60–120 years, the minimum of 10–15 years to reach reproductive maturity, and the sporadic seed production provide obstacles to breeding Norway spruce. The traits given particular attention by breeders are: (1) high volume production combined with a growth rhythm that allows timely cold acclimation in the fall and avoidance of damage from late frosts in the spring, (2) quality characters such as straightness, branch thickness and angle, wood density, fibre quality, cellulose content and lignin content and composition, and (3) reduced susceptibility to insects and pathogens, particularly the root-rotting fungi such as *Heterobasidion annosum*. Shortening the juvenile phase is of special interest for accelerating the breeding programme.

It is now possible to produce transgenic plants of Norway spruce routinely, both by *Agrobacterium*-mediated transformation (Wenck et al. 1999; Klimaszewska et al. 2001) and by particle bombardment (Walter et al. 1999; Clapham et al. 2000a, b). This opens the way to many research and practical applications. The main limitations at present are set not so much by technical problems as by restraints arising from environmental considerations, particularly the demands for certification by the Forest Stewardship Council (Strauss et al. 2001; Fenning and Gershenzon 2002).

8.2 Types of Particle Accelerator

Particle bombardment, i.e., microprojectile bombardment, is based on the acceleration of high-velocity "microprojectiles" (usually gold particles with a

Molecular Methods of Plant Analysis, Vol. 23
Genetic Transformation of Plants

diameter between 0.5 and 3.0 μm, formerly tungsten particles) coated with DNA into intact plant cells (Klein et al. 1987). The types of accelerator ("gun") that have been used for bombardment of Norway spruce cells include the original commercial machine after Klein et al. (1987), PDS-1000, supplied by DuPont, Wilmington, DE, firing modified cartridges (Robertson et al. 1992); its commercial successor, PDS-1000/HE (Bio-Rad), powered by a burst of helium gas (Robertson et al. 1992; Walter et al. 1999); a custom-made electric discharge accelerator modified by Dr. J. Watts from the design by Christou et al. (Newton et al. 1992); and a custom-made "particle inflow" gun modified by Prof. H.-U. Koop from the design by Finer et al. (1992), in which a drop of a suspension of gold particles coated with DNA is projected directly into the cells with a burst of helium gas (Clapham et al. 2000a). In the Bio-Rad accelerator, the suspension of gold particles is dried out on a small circle of light plastic, the "macroprojectile"; the macroprojectile is forced by the burst of helium gas against a circular net (stopping plate), the gold particle microprojectiles continuing through the holes of the stopping plate into the cells. Transgenic Norway spruce plants have been produced at comparable frequencies by the particle inflow gun and the Bio-Rad accelerator. The particle inflow gun is no more expensive than an electrophoresis apparatus and has much lower running costs than the Bio-Rad accelerator. The Bio-Rad accelerator is expensive, but is higher powered and is more suitable for studies of transient expression in some cell types.

8.3 Transformation of Embryogenic Cultures

8.3.1 Transient Expression in Embryogenic Cultures

Particle bombardment is usually the best method for delivering DNA to cells for assays of promoter activity by transient expression (Potrykus 1991). Early studies of transient expression in conifer material (reviewed Ellis 1995; Clapham et al. 2000b) were often directed at finding suitable promoters, preferably inducible, for driving selectable markers. Various promoters fused to the *gusA* (i.e., *uidA*) reporter gene coding for β-glucuronidase were shot into embryogenic cultures and β-glucuronidase activity measured 1–2 days after bombardment. This was done either by counting blue foci after histochemical staining or by using the fluorometric assay (Jefferson 1987). To compare the activity of various promoters, the most efficient procedure is to incorporate an internal control. For example, Clapham et al. (2000a) coated the gold particles with both a plasmid containing a promoter-*gusA* construct and a control plasmid containing the *luc* gene coding for luciferase under an enhanced 35S promoter. The use of the internal control enables a correction to be made for shot-to-shot variation (e.g., Martinussen et al. 1994; Clapham et al. 2000a) and reduces the amount of replication required to detect significant differences in promoter activity. In embryogenic cultures of Norway spruce, the maize and the sunflower ubiquitin promoters are both far more strongly expressed than the standard unduplicated 35S promoter, which is relatively weakly expressed (Yibrah et al. 1994; Clapham et al. 2000a). To date, a really useful inducible promoter has not been shown to work in Norway spruce, or any other conifer to our knowledge, although several promoters show varying degrees of inducibility (reviewed Ellis 1995; Clapham et al. 2000b). A copper-inducible system has worked promisingly in three angiosperm systems (e.g., *Arabidopsis*, Granger and Cyr 2001), and is worth investigating in conifers, despite potential problems with copper toxicity and induction of nontarget genes.

Transient expression (and stable transformation) in embryogenic cultures of Norway spruce (Yibrah et al. 1994; Walter et al. 1999; Clapham et al. 2000a) was greatly enhanced by exposing the cells to high osmoticum, such as 0.125–0.25 M inositol or sorbitol, for an hour or more before bombardment. This confirms earlier work with tobacco suspension cultures (Russel et al. 1992). The time of bombardment after subculture affected levels of transgenic expression, with an optimum at 3–5 days for three different rapidly growing cell lines (Clapham, unpubl.).

8.3.2 Production of Stably Transformed Cell Cultures and Transgenic Plants

The procedures for the routine production of transgenic Norway spruce plants by particle bombardment (Walter et al. 1999; Clapham et al. 2000a) differ somewhat in details. Briefly, Walter and coworkers maintained the embryogenic cultures on solid medium. About 16–20 h before transformation, embryogenic tissue (250 mg) was suspended in 1 ml liquid medium and spread onto a sterile filter paper placed on solid culture medium containing 0.25 M sorbitol. The tissue was left overnight in a closed but unsealed Petri dish in a laminar flow hood to dry. Gold particles (1.5–3.0 μm) were coated with a plasmid pCW122 containing the *nptII* gene coding for resistance to kanamycin and geneticin driven by the CaMV 35S promoter and the *gusA* reporter gene driven by a double 35S promoter. The cells were bombarded with the PDS-1000He device (BioRad Inc.). Three days after bombardment, the filter papers were transferred onto solid culture medium containing 5 μM 2,4-D, 2.5 μM benzylaminopurine and 15–20 mg/l of geneticin for selection of stably transformed cells. The cultures were maintained at 24°C under low light conditions (5 $\mu mol\, m^{-2} s^{-1}$) with subculturing every 2 weeks. Plantlets were regenerated by a procedure beginning with embryo maturation on medium containing abscisic acid (ABA; for details see Walter et al. 1999).

The Uppsala group worked with embryogenic suspension cultures, usually newly started from thawed cryogenically preserved lines stored in liquid nitrogen. The cultures were maintained by weekly subculture in proliferation medium (modified half-strength LP medium containing 2,4-D (9 μM), benzylaminopurine (4.4 μM), 1% sucrose and, for most cell lines, 450 mg/l glutamine;

(cell line A66 used for some of the work described in Clapham et al. (2000a) did not require glutamine, however). Cells were grown in the dark at 22°C. An advantage of using suspension cultures is that large quantities of cells are easily obtained. Cells were taken for bombardment 4–5 days after subculture. One to 3h before bombardment, the cells were allowed to sediment in 50-ml Falcon tubes and the medium was changed to half-strength LP medium as before, but without glutamine and containing 0.125–0.25M inositol as osmoticum. Under these conditions the cells were plasmolyzed. Cell samples of 100–150mg fresh weight were collected under mild vacuum onto filter paper disks of diameter 4cm, which were placed on 12-cm-diameter filter papers, soaked in the same medium, in 15-cm Petri dishes. Gold particles, 1.5–3.0μm diameter, were coated using precipitation with sodium acetate and alcohol with the plasmid pAHC25 (Christensen and Quail 1996). The plasmid contains the *bar* gene coding for resistance to the herbicide Basta and the *gusA* reporter gene, each driven by a maize ubiquitin promoter.

After bombardment on day 0, using the modified particle inflow gun mentioned earlier (Sect. 8.2), filter papers with the cells were placed on proliferation medium supplemented with 0.25M inositol and solidified with gelrite (0.18%). On day 8, cells were subcultured onto selection medium, which was the previous medium supplemented with Basta (to give 0.2mg/l ammonium glufosinate) and 5-azacytidine (3mg/l) On day 15, the cells were subcultured on the same medium without 5-azacytidine and were subcultured on the same medium monthly. Embryogenic tissue resistant to Basta appeared after 2–4 months. The resistant colonies were allowed to proliferate on standard proliferation medium without raised inositol content. Somatic embryos were matured on medium containing ABA (Bozhkov and von Arnold 1998), then subjected to drying at high humidity. Plantlets were regenerated and grown on in the greenhouse (Fig. 8.1). Transformation was confirmed, as in the work of Walter et al. (1999), by PCR, by Southern blotting and by β-glucuronidase activity. Of 11 transformed sublines examined by Southern blotting, 4 contained the transgenes in low copy number (1–3), the rest in copy numbers of 15–20.

In subsequent work the procedure has been modified in some respects. Many embryogenic lines of Norway spruce, e.g., lines 95:88:22 and 95:61:21, require reduced nitrogen for rapid proliferation. This is usually provided as glutamine, which interferes with the selection of transformants resistant to Basta on selective media (Clapham, unpubl. observ.). Basta inhibits the enzyme glutamine synthetase, which is thought to be required not only for the synthesis of glutamine, but also for the removal of toxic ammonium ions produced by various normal metabolic reactions (e.g., Oaks and Hirel 1985), so that it is not self-evident that Basta would fail to work effectively in the presence of glutamine. Few transformants are obtained with the above-mentioned lines if no source of reduced nitrogen is provided in the selective medium. We have found, however that the incorporation of asparagine (150mg/l; Bozhkov et al. 1993) into the selective medium allows the recovery of transformants from these lines at rates comparable to those originally obtained with line A66. We

Fig. 8.1. Transgenic Norway spruce plants in the greenhouse, 3–4 years after transformation

presume that the asparagine is broken down to aspartic acid by an asparaginase faster than it is converted to glutamine by the reverse action of an asparagine synthetase (glutamine-hydrolyzing) enzyme (e.g., Lam et al. 1998).

8.3.3 Stability of Transgene Expression

An important question, particularly if transgenic Norway spruce plants are to be used in practical forestry, is whether the transgene is reliably expressed over the relevant time period, which may be as long as 60–120 years in Scandinavia. Epigenetic transgene silencing has often been noted; review, Matzke et al. (2002). We have tested the stability of expression of the *bar* gene conferring Basta resistance driven by the maize ubiquitin promoter over 36–39 months; see Fig. 8.2 and legend.

8.3.4 Trends in Transgenic Plant Production

Now that it is possible to produce transgenic plants of Norway spruce routinely by *Agrobacterium*-mediated transformation (Wenck et al. 1999; Klimaszewska

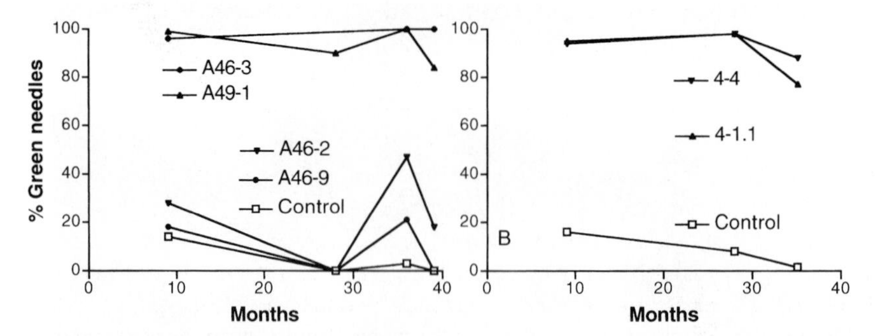

Fig. 8.2. Stability of *bar* gene expression driven by the maize ubiquitin promoter in plants regenerated from various sublines of transgenic Norway spruce over 36–39 months. Percentage of needles showing resistance to 500 mg/l Basta; for details of method, see Brukhin et al. (2000). At 9 months, the plants were in a growth room; at 28 months, in June, in the greenhouse after budburst; at 36 months, in February, in the greenhouse before budburst; at 39 months, in May in the greenhouse after budburst. *Left* Untransformed control line 95:88:22 and four sublines transformed with *bar*. *Right* Untransformed control line 95:61:21 and two sublines transformed with *bar*. The degree of resistance to Basta has remained fairly stable above the controls

et al. 2001), as well as by particle bombardment, the special features of each method should be considered. To date, there is no controlled study for Norway spruce directly comparing each method. Usually, agrobacterial methods are regarded as potentially more efficient, producing transgenic plants at a higher frequency, at least for selected genotypes. Furthermore, a higher proportion of the plants usually contains the transgene in a single unmodified copy, and the transgene is less susceptible to silencing. Particle bombardment is considered less genotype-specific. This holds broadly true for rice, one of the few species for which controlled comparisons have been made. A *japonica* rice variety was transformed by each method with a construct carrying three marker genes (Dai et al. 2001). The average gene copy number was 1.8 by *Agrobacterium* and 2.7 by particle bombardment, not a vast difference; expression from the *gusA* reporter gene was on average slightly higher, less variable and more stable over generations with *Agrobacterium*; and fertility was higher with *Agrobacterium* (about twice the seed-set). Particle bombardment was described as "a high efficiency system to produce a large number of transgenic plants with a wide range of gene expression", rice being, like conifers, somewhat recalcitrant to *Agrobacterium*. An important difference between the methods, observed also in this study, is that with particle bombardment, multiple copies of the transgenes tend to be integrated at the same site, explained by the two-phase integration mechanism of Kohli et al. (1998). Multiple copies tended to be at different genomic sites with *Agrobacterium*.

Recent studies have shown how to reduce the copy number, and consequential gene silencing, with particle bombardment. Rice tissue was bombarded with a whole plasmid, supercoiled or linearized, containing the *bar* gene, or with a linear DNA fragment containing a minimal *bar* gene cassette. Southern blot analysis showed striking differences in the integration pattern with a bombarded construct. As much as 77% of the plants from bombardment with the minimal construct showed a single band, indicating a single site and usually low copy integration, whereas the plants from bombardment with whole plasmids, supercoiled or linearized, showed complex patterns (Fu et al. 2000). For a method of obtaining single copy transformants using a recombinase construct, see Srivastava and Ow (2001).

Particle bombardment with minimal gene cassettes rather than plasmids also rules out the possibility of undesired vector sequences being incorporated into the transgenic plant, as happens at high frequency with *Agrobacterium* (for references, see Fu et al. 2000). Another advantage of particle bombardment is the ease with which plants can be transformed with several genes by coating the gold particles with up to 12–13 constructs (for soybean, Hadi et al. 1996; for rice, Chen et al. 1998). In Norway spruce, also, high frequencies of cotransformation of genes carried on different plasmids have been found (Clapham et al. 2000a; Bishop-Hurley et al. 2001). From a regulatory point of view, especially with long-lived plants, it is an advantage that there is no possibility that the potentially pathogenic *Agrobacterium* remains in the plants. Particle bombardment will likely continue to be attractive for Norway spruce and other conifers, despite the availability of agrobacterial methods of transformation.

8.4 Transformation of Pollen

The use of pollen, the natural genetic carrier, as a vector for genetic transformation of conifers has several advantages. Pollen lots can readily be collected from a large number of trees, they are easy to store for long periods of time, and crosses can be made with several mother trees to produce highly divergent transgenic plants. Pollen lots are composed of numerous genotypes, each pollen grain being genetically unique. After pollen transformation, any transgenic progeny will result from natural processes. Selectable marker genes are not essential, though their use might improve the efficiency of the production of transformants and decrease the amount of work in progeny screening. The transgenic progeny can be propagated vegetatively via somatic embryogenesis.

8.4.1 The Reproductive Biology of Norway Spruce

Norway spruce is monoecious with both male and female strobili on the same individuals. The reproductive buds are initiated during the growing season of the year before. Male meiosis and pollen development occur after the winter dormancy in late winter or early spring and they are regulated by the temperature conditions (Luomajoki 1993). Female meiosis occurs close to or during the period of pollination (Sarvas 1968). The pollen chamber can accommodate, on the average, five pollen grains, thus providing an opportunity for male gametophyte competition through differential germination ability and tube growth rate (Nikkanen et al. 2000; Aronen et al. 2002).

A male flower of Norway spruce may contain as many as 600,000 pollen grains, which are released during anthesis and dispersed by wind over long distances. A mature pollen grain is approximately 100 μm in diameter and is composed of five cells: two partially degenerated prothallial cells embedded in the intine, one large tube cell, a stalk cell and a body cell (Martinussen et al. 1994). Dried pollen can be stored over prolonged periods without loss of germination capability (Lanteri et al. 1993; Hak 1996).

8.4.2 Transient Expression in Pollen

The use of pollen, the natural genetic carrier, as a vector for genetic transformation of conifers has several advantages. Pollen lots can readily be collected from a large number of trees, they are easy to store for long periods of time, and crosses can be made with several mother trees to produce highly divergent transgenic plants. Pollen lots are composed of numerous genotypes, each pollen grain being genetically unique. After pollen transformation, any transgenic progeny will result from relatively natural processes, without a tissue culture phase and the problems this often entails. To reduce propagation costs, however, the procedure may be combined with vegetative propagation techniques, such as somatic embryogenesis or cutting production or both. Last, but not least, selectable marker genes are not essential, though their use might improve the efficiency of the production of transformants and decrease the amount of work in progeny screening.

Particle bombardment of microspores or microspore-derived tissues has been used successfully to achieve transient expression and production of transgenic plants, for instance, in several angiosperm species, e.g., tobacco (Stöger et al. 1995). Several reports describe transient expression of reporter genes in various gymnosperm species (Hay et al. 1994; Li et al. 1994), including Norway spruce (Martinussen et al. 1994, 1995; Häggman et al. 1997). Regeneration of transgenic plants by applying transformed pollen in conventional pollination has, however, been less common. There is a published report for tobacco (van der Leede-Plegt et al. 1995), unpublished information for white spruce (D.D. Ellis. pers comm., see Clapham et al. 2000b), but otherwise in gymnosperms techniques of using bombarded pollen with conventional pollination have been published only for Scots pine (*Pinus sylvestris* L.) and Norway spruce (Aronen et al. 1998).

Generally, in pollen transformation the selection of transformation technique is of key importance because nucleases associated with microspores/ pollen may especially complicate chemically and electrically induced gene delivery (Martinussen 1994; Vischi and Marchetti 1997). Clonal differences in competence for genetic transformation also play an important role in pollen transformation. All pollen sources are not equally appropriate for gene transfer (Häggman et al. 1997), though unsuitable pollen donors may be used as mother trees in controlled pollinations with bombarded pollen. Coniferous pollen is multicellular, consisting of generative and vegetative cells and generative cells must be transformed to obtain transgenic progenies. Moreover, the fact that in Norway spruce, as in many coniferous species, the pollen chamber can accommodate several pollen grains, means that transgenic and nontransgenic grains compete to achieve fertilization.

In pollen transformation of Norway spruce, the spectrum of pollen genotypes, as reflected in the pollen origin, as well as the regulatory sequences of the reporter gene, had significant effects on expression. The highest β-glucuronidase (GUS) expression (44 ± 12% of germinated pollen grains) was achieved with an abscisic acid-inducible promoter (Häggman et al. 1997), indicating a good starting point for obtaining transgenic progenies. Out of the several bombardment pressures tested between 400 and 1300 psi no significant differences could be found, but when the pollen samples were bombarded twice significantly higher results were achieved than with single bombardments (Häggman et al. 1997). When GUS-expressing pollen grains as a percentage of germinated grains were followed during the first 3 days of pollen germination in vitro, GUS expression rapidly decreased (Häggman et al. 1997). Thus, it was an open question whether in some cases the transgenes were stably integrated into the genomes of the pollen cells (generative or vegetative), or whether all the expression should be considered transient.

8.4.3 Development of Controlled Pollination Techniques for Bombarded Pollen

Despite numerous reports on transient transformation of pollen in various species, regeneration of transgenic plants after conventional pollination with transformed pollen has been rare. In gymnosperms, methods for the dehydration and storage of transformed pollen compatible with conventional crossing techniques have been developed only for Scots pine and Norway spruce (Aronen et al. 1998).

In gymnosperms controlled pollinations are usually performed by applying dried pollen into isolation bags with female strobili. This technique has been successful and generally used in Norway spruce also even if seed yield is sometimes lower than with open pollination (Alfjorden 1978). However, because preparation of pollen suspensions is necessary for particle bombardment (Häggman et al. 1997; Aronen et al. 1998) the first tested technique for controlled pollinations was so-called liquid pollination originally developed for radiata pine (*Pinus radiata* D. Don) by Sweet and collaborators (1992, 1993). The percentage of mature cones and full seed was, however, lower with liquid than with conventional pollinations. To increase the chances of obtaining transgenic progenies, methods for dehydration of bombarded pollen were studied (Aronen et al. 1998). The best method was to dry the pollen suspension in a desiccator at room temperature for 24h. Long-term storage of bombarded pollen lots at –20°C was also successful (Aronen et al. 1998).

Stably transformed coniferous seedlings obtained through the application of transformed pollen in controlled crossings has thus far been demonstrated only in the case of Scots pine (Aronen et al. 2003). Simultaneous work for production of transgenic Norway spruce seedlings is, however, going on with some positive results at PCR-level. Progress with Norway spruce has been slow owing to the irregular flowering and the sensitivity of seed development to weather conditions.

Potentially transgenic conifer seedlings have, to date, been produced without selection, which has required laborious screening. Candidate seedlings have been selected based on histochemical β-glucuronidase assay followed by DNA isolation, PCR amplification and then Southern analysis (Aronen et al. 2003). The final frequencies of transgenic progenies obtained are low (Aronen et al. 2003). Possible reasons are: integration of the transgene into the nucleus of a vegetative rather than generative cell; nonoptimal histochemical β-glucuronidase assay (recently applied acetone pretreatments of the assayed samples have been found to be necessary for strong GUS expression); or nonoptimal sampling times. The fact that the developing conelets have to be covered by plastic net bags to prevent the release of potentially transgenic material into the environment may have mechanically destroyed some of the developing strobili and contributed to the decrease in the number of candidate seedlings. Stably transformed Norway spruce seedlings via the application of pollen transformation and controlled crossing are, however, likely in the near future.

8.5 Applications of Transgenic Norway Spruce in Research

8.5.1 Genes Regulating Embryogenesis

Combined with cryopreservation, somatic embryogenesis is an attractive method of vegetative propagation in many species including Norway spruce, both as a tool in breeding programmes and for large-scale clonal propagation of elite material (e.g., Högberg et al. 1998, 2001; von Arnold et al. 2002). Further, as indicated above, somatic embryos or embryogenic cultures are, at present, the usual starting material for production of transgenic Norway spruce. Therefore, there is much interest in working out the cell and molecular biology of how somatic embryogenesis is regulated, in Norway spruce and other species.

Pa18 is a Norway spruce gene encoding a protein resembling a lipid transfer protein (Sabala et al. 2000). From in situ hybridization studies (Sabala et al.

2000), it was found that *Pa18* is equally expressed in all cells of proliferating embryogenic cultures (proembryogenic masses), but during embryo maturation the expression is stronger in the outer cell layer. This outer layer is regarded as similar to the protoderm of angiosperm embryos. To further study the action of *Pa18* on embryogenesis, embryogenic cultures were bombarded with gold particles coated with a plasmid containing the *bar* gene fused to the maize ubiquitin promoter as selectable marker and the *Pa18* cDNA fused to another copy of the maize ubiquitin promoter. Twenty independent transgenic sublines ectopically expressing *Pa18* were examined. Maturation of the embryos in response to abscisic acid was disturbed in the transformed cell lines. This was attributed to the constitutive expression of *Pa18* in all cells of the developing transgenic embryos in the overexpressing lines, in contrast to the more localized expression in the outer cell layer of the untransformed cultures. The cells of the outer layer in the transformants were often elongated and vacuolated instead of remaining small and uniform, and the embryos lacked the smooth surface of untransformed embryos. The observations were extended in Hjortswang et al. (2002).

A homeobox gene, *PaHB1*, (for *Picea abies* homeobox1) similarly switches from ubiquitious expression in proembryogenic masses to expression limited to the outer cell layer in maturing embryos (Ingouff et al. 2001). Again, ectopic expression of the gene directed by the maize ubiquitin promoter leads to an early developmental block. Transformed embryos lack a smooth surface.

A gene isolated from loblolly pine, *PtNIP1;1*, encodes an aquaglyceroporin that is expressed early in embryogenesis and preferentially in the suspensor (Ciavatta et al. 2001). The promoter region (899 bp upstream of the open reading frame) was fused to the reporter gene *gusA* (Ciavatta et al. 2002). The resulting plasmid, pNIP-GUS, mixed with modified pAHC25 lacking the GUS cassette, was coated onto gold particles and bombarded into embryogenic Norway spruce cells. Three transgenic sublines were examined by histochemical GUS assays (Ciavatta et al. 2002). GUS activity was strong in all cells of proembryogenic masses. During a brief period of early embryogeny on maturation medium, GUS staining was intense in the suspensor region, but not detectable in embryonal masses. Mature embryos showed no GUS expression. In contrast, a control transgenic line transformed with the *gusA* gene fused to a *Eucalyptus* cinnamylalcohol dehydrogenase gene promoter expressed GUS throughout embryo development.

8.5.2 Genes with Similarity to Defense Genes

Heterobasidion annosum is a major pathogen on Norway spruce. The fungus can infect living conifer roots of all ages and causes root and butt rot (Woodward et al. 1998). Genes modifying susceptibility to the disease are of great interest. In a screening of a cDNA library made from *Pythium dimorphum*-infected Norway spruce seedling roots, two cDNA clones showing simi-

Table 8.1. Vertical spread of *Heterobasidion annosum* in the sapwood of 8-month-old *spi-1* transformed Norway spruce. Two separate infection experiments (A and B) were carried out on 8-month-old Norway spruce plants. Plants were inoculated by attaching a *H. annosum*-infected spruce dowel to a freshly made 5-mm circular wound. Fungal growth was scored 34 days after inoculation with *H. annosum*. Each stem was cut in 5-mm-thick disks which were placed on moist filter paper in a Petri dish. The disks were examined for the presence of conidiophores on the surface 10 days after sampling. In experiment A, plants from sublines 4–1.0 and 4–3 as well as the untransformed control line were infected. In experiment B, plants from sublines 4–1.1 and 4–4 and the untransformed control line were infected. The data are presented as a cumulative mean and the data from each experiment were analysed separately using the Mann-Whitney U-test. There was a significant reduction in fungal growth in subline 4–4 ($P < 0.01$)

A			B		
Subline	Number of individuals	Fungal growth (mm)	Subline	Number of individuals	Fungal growth (mm)
Control	6	78	Control	15	92
4–1.0	17	71	4–1.1	18	103
4–3	9	73	4–4	19	72[a]

[a] $P = 0.01$.

larity to plant defensins and peroxidases were isolated. The clones were designated *spi1* and *spi2* respectively (Sharma and Lönneborg 1996; Fossdal et al. 2001). Plant defensins and plant peroxidases from angiosperms have been classified as pathogenesis-related (PR) proteins (van Loon and van Strien 1999).

To evaluate the effect of the SPI 1 protein on Norway spruce defence, the cDNA was fused to an enhanced CaMV 35S promoter and transferred to a plasmid also harbouring the *bar* gene fused to the maize ubiquitin promoter. Embryogenic cultures were bombarded with gold particles coated with the resulting plasmid and 14 independent transgenic sublines were isolated (Elfstrand et al. 2001a).

Inoculation of 8–10 month old transgenic plants with *H.annosum* showed that only the plants from the transgenic subline with the highest accumulation of SPI 1 protein, two to threefold as judged by western analysis, showed reduced growth of the fungus in the sapwood (Table 8.1; Elfstrand et al. 2001a). This observation is in line with other reports that the accumulation of plant defensin needs to reach a threshold level in transgenic plants to improve protection against pathogens (Terras et al. 1995; Gao et al. 2000).

Plant peroxidases catalyse many processes in plant cells. For instance, reinforcement of the cell wall (Lagrimimi et al. 1993), oxidation of phenolics (Bradley et al. 1992; Kobayashi et al. 1994) and production of toxic radicals (Bolwell and Wojtaszek 1997) can be involved in plant defence. To study the physiological role of *spi 2* in Norway spruce, a transformation vector with the *spi 2* gene was fused to the maize ubiquitin promoter and the *bar* gene fused to another copy of the maize ubiquitin promoter was constructed. Thirty independent transgenic sublines were isolated after bombarding embryogenic tissues with the resulting plasmid. The isolated sublines exhibited up to 40-fold increased peroxidase activities. Furthermore, sublines could be separated into a class I-type of sublines, showing no alteration in the total peroxidase activity compared to the control and into a class II-type of sublines with significantly increased total peroxidase activity. Compared to the control and class I sublines, the class II-type of sublines showed a reduced epicotyl to root formation ratio during germination, reduced survival at 5 months and reduced height growth (Fig. 8.3). Histochemical assays for peroxidase activity and lignin (phloroglucinol: detecting primarily coniferaldehyde) showed more intense staining in the class II-type of sublines than in the control, although there was no increment in the Klason lignin (Elfstrand et al. 2001b).

Ectopic expression of *spi 2* under control of the CaMV 35S promoter in tobacco resulted in decreased growth of the bacterial pathogen *Erwinia carotovora* in the tobacco plants. The analysis of the *spi 2*-expressing tobacco lines suggested that although there was no change in the lignin content, in the tobacco stems, lignin composition was changed. The relative frequency of coniferaldehyde in (β-O-4-linked G type of lignin subunits) was slightly increased. The increased coniferaldehyde frequencies were perhaps the result of altered redox-status in the plant cell walls (Elfstrand et al. 2002), so that coniferyl alcohol was reduced to coniferaldehyde.

Spi 1- and *spi 2*-transformed Norway spruce plants are being grown in a glass-house to allow further testing of the effect of *spi 1* and *spi 2* on Norway spruce disease resistance. The *spi 1* lines are similar to the control line after 4 years of cultivation while the 3-year-old *spi 2* transformed lines still exhibit alterations in shoot growth (Fig. 8.3).

The studies with *spi 1* and *spi 2* show that production of transgenic Norway spruce plants expressing genes of interest under the control of constitutive promoters can be a useful method to test candidate genes for molecular breeding and to evaluate the physiological role of a candidate gene in Norway spruce.

8.6 Prospects for Transgenic Norway Spruce in Practical Forestry

Two aspects will be considered briefly here, the economic and the ecological; for the stability of gene expression, see Sect. 8.3.3 above. In order to be economically justified in practical silviculture, the actual benefits of genetically engineered forest trees should exceed the costs of their production. For Norway spruce, the interval between establishment and harvest is usually long, around 60–80 years. Improvement in quality traits will not be realized until the better wood is actually harvested, while improvement in adaptation and growth can have an impact on the allowable cut from a forest estate upon establishment. When rotation is long, the discounted present value of even a small increase in annual growth rate can look more attractive than an improvement in fiber

Fig. 8.3A—C Growth of transgenic Norway spruce overexpressing **spi 1** (encoding a defensin-like protein) or **spi 2** (encoding a cationic peroxidase). Four-year-old **spi 1**-transformed and 3-year-old **spi 2**-transformed Norway spruce plants, with leader shortened to encourage bushy growth for propagation by cuttings. **A** The *spi 1* plants are of normal appearance. The **spi 2**-transformed plants showed significantly reduced growth during their first season in climate chambers with increased sensitivity to changes in the conditions; they continued to show reduced growth with clearly delayed bud break and flush (**B**) as compared to the untransformed control plants (**C**)

quality that cannot be realized until harvest. This reflects the fact that investments and management decisions in forestry are usually made based on volume-driven thinking (Mullin and Bertrand 1998). Taking into account these economical aspects as well as the long rotation cycle of the Scandinavian conifers, target traits for genetic engineering should be carefully chosen in order to gain real and sustainable benefit for both forest owners and the whole society.

Forest ecosystems have several unique characteristics. Human input to forested areas is smaller, and species-level biodiversity is higher, than with agriculture. The interactions among forest species are complex, and although this complexity of the forest ecosystem is well recognised, it is not properly understood. The maintenance of diversity for nonharvested species is, however, nowadays usually considered as an essential element of sustainable forest management (Mullin and Bertrand 1998). The above-mentioned characteristics are especially important in Norway spruce and Scots pine stands in Scandinavia, these being mainly multiple-use forests and in addition to their use for timber production are also important facilities for recreation, collecting of wild berries, edible mushrooms and lichen, hunting, and reindeer husbandry (Sevola 1998). In fact, regardless of their ownership, forests in Scandinavia are often seen as national property and, therefore, the legislation and guidelines for their management need to achieve both political and social approval.

There is, of course, a trade-off between biodiversity and production. The maximum sustainable production of timber from a natural forest anywhere in the world may be as low as $2\,m^3ha^{-1}year^{-1}$, whereas a plantation of *Pinus radiata* in New Zealand produces on average $20\,m^3ha^{-1}\,year^{-1}$ (Fenning and Gershenzon 2002). If biodiversity in Scandinavian forests is maximized by changing from plantations to naturally regenerating "autochthonous" forests, the low yields may make it uneconomic to cut down the trees, as appears to have happened with Norway spruce in Switzerland (Ortloff 1999). Transgenic trees will be grown in plantations. For an important if controversial discussion of environmental issues, including a section on transgenic crops, see Lomborg (2001).

There is substantial concern that transgenic trees can produce unforeseen negative consequences for managed and natural ecosystems. A comprehensive model for risk assessment related to genetic modification of trees includes both identification and evaluation of biological risks and assessment of public perceptions regarding these risks.

The following questions need careful attention.

1. What is the risk?
2. How likely is the risk to occur?
3. What is the severity and extent of the effect if it occurs?
4. What are the expected benefits?
5. Is the risk acceptable?
6. How important is risk assessment for different factors?
7. What means are available for risk management?

From an ethical perspective, it is important to note that a society free of risks is not possible because vital values in terms of biodiversity, development in third-world countries, better and new qualities of wood and increased carbon sequestration are at stake, and these values have to be balanced against risks of ecological damage. This implies that the approach to risk assessment should be normalization through comparison of gene-technology-associated risks with the risks of conventional and future technologies for breeding and development in forestry.

In nature, a new allele will typically appear as a rare variant. New alleles can enter the population by migration (pollen and/or seed dispersal) or mutation. It can have various fates, which range from rapid extinction, through persistence at low frequencies, to spreading through the population more or less slowly and ultimately become fixed. The fate will depend on whether it is selectively neutral or significantly advantageous. If the gene is deleterious, it will be eliminated or remain at a low frequency with minimal cost in population fitness. Wind-pollination gives potential for long-distance spread of genetic material into native stands. While most of the pollen settles within a short distance from points of release, dense pollen clouds can still occur at considerable distance from large areas of stands (Lindgren et al. 1995).

The result of a genetic transformation is an organism that carries a new sequence of DNA at one or more locations within the genome. If such a sequence is already present in nature, we have something very much like the possible outcome of natural gene transfer. The use of transgenics can immediately raise the effective frequency of a new gene to 100% in a plantation. There are various categories of transgenes. Constructs that modify the expression of endogenous genes or switch off expression of endogenous genes are likely to present relatively low risks. Another category represents structural genes of a novel function within the species. For these novel genes the attendant risk might be higher.

Suppression of reproduction is an attractive goal, offering up to 100% reduction in pollen flow. Genetic engineering strategies exist to achieve this goal, including expression of cytotoxic genes in reproductive tissue, repression of the reproductive pathway through homeotic genes and suppression of genes involved in reproductive development. None of these methods can guarantee complete sterility (e.g., Strauss et al. 1995) and particularly for long-lived conifers, it is preferable to ensure that the spread of pollen/seeds will not have negative effects on the environment.

Gene flow is a prerequisite for many potential ecological impacts of transgenic organisms. Some level of gene flow is likely for transgenic trees with interfertile wild relatives. Therefore, an in-depth analysis of the dynamics and potential for gene flow from plantations will form an essential part of the risk assessments for a wide variety of traits and environments. Methods are needed to extrapolate from small-scale, short-term studies to appropriate temporal and spatial scales to allow transgenic risk assessments for trees. These have been developed by Strauss and his co-workers at Oregon State University. A model, together with a user's manual, simulating transgenic effects in a variable environment, originally designed for transgenic poplar, is available at the website http://www.fsl.orst.edu/tgerc/STEVE_model/.

References

Alfjorden G (1978) Experiment with various isolation materials for controlled pollination with spruce. Association for Forest Tree Breeding, Institute of Forest Improvement, Uppsala, Sweden, Yearbook 1977, pp 72–78

Aronen T, Nikkanen T, Häggman H (1998) Compatibility of different pollination techniques with microprojectile bombardment of Norway spruce and Scots pine pollen. Can J For Res 28:79–86

Aronen T, Nikkanen T, Harju A, Tiimonen H, Häggman H (2002) Pollen competition and seed-siring success in *Picea abies*. Theor Appl Genet 104:638–642

Aronen T, Nikkanen T, Häggman H (2003) The production of transgenic Scots pine (*Pinus sylvestris* L.) via the application of transformed pollen in controlled crossings. Transgen Res (accepted)

Bishop-Hurley SL, Zabkiewicz RJ, Grace L, Gardner RC, Wagner A, Walter C (2001) Conifer genetic engineering: trasngenic *Pinus radiata* (D.Don) and *Picea abies* (Karst) plants are resistant to the herbicide Buster. Plant Cell Rep 20:235–243

Bolwell GP, Wojtaszek P (1997) Mechanism for the generation of reactive oxygen species in plant defence – a broad perspective. Physiol Mol Plant Pathol 51:347–366

Bozhkov PV, von Arnold S (1998) Polyethylene glycol promotes maturation but inhibits further development of *Picea abies* somatic embryos. Physiol Plant 104:211–224

Bozhkov PV, Mikhlina SB, Shiryaeva GA, Lebedenko LA (1993) Influence of nitrogen balance of culture medium on Norway spruce {*Picea abies* (L.) Karst} somatic polyembryogenesis: high frequency establishment of embryonal-suspensor mass lines from mature zygotic embryos. J Plant Physiol 142:735–741

Bradley D, Kjellborn P, Lamb C (1992) Elicitor- and wound-induced oxidative cross-linking of a proline-rich plant cell wall protein: a novel, rapid defense response. Cell 70:21–30

Brukhin V, Clapham D, Elfstrand M, von Arnold S (2000) Basta tolerance as a selectable and screening marker for transgenic plants of Norway spruce. Plant Cell Rep 19:899–903

Chen L, Marmey P, Taylor NJ, Brizard J-P, Espinoza C, D'Cruz P, Huet H, Zhang S, de Kochko A, Beachy RN, Fauquet CM (1998) Expression and inheritance of multiple transgenes in rice plants. Nature Biotechnol 16:1060–1064

Christensen AH, Quail PH (1996) Ubiquitin promoter-based vectors for high-level expression of selectable and/or screening marker genes in monocotyledenous plants. Transgen Res 5:213–218

Christou P (1990) Soybean transformation by electric discharge particle acceleration. Physiol Plant 79:210–212

Ciavatta VT, Morillon R, Pullman GS, Chrispeels MJ, Cairney J (2001) An aquaglyceroporin is abundantly expressed early in the development of the suspensor and the embryo proper of loblolly pine. Plant Physiol 127:1556–1567

Ciavatta V, Egertsdotter U, Clapham D, von Arnold S, Cairney J (2002) A promoter from the loblolly pine PtNIP;1 gene directs expression in an early-embryogenesis and suspensor-specific fashion. Planta 215:694–698

Clapham DH, Demel P, Elfstrand M, Koop H-U, Sabala I, von Arnold S (2000a) Gene transfer by particle bombardment to embryogenic cultures of *Picea abies* and the production of transgenic plantlets. Scand J For Res 15:151–160

Clapham DH, Newton RJ, Sen S, von Arnold S (2000b) Transformation of *Picea* species. In: Jain SM, Minocha S (eds) Molecular biology of woody plants. Kluwer, Dordrecht, pp 105–118

Dai S, Zheng P, Marmey P, Zhang S, Tian W, Chen S, Beachy RN, Fauquet C (2001) Comparative analysis of transgenic rice plants obtained by *Agrobacterium*-mediated transformation and particle bombardment. Mol Breed 7:25–33

Elfstrand M, Fossdal C-G, Swedjemark G, Clapham D, Olsson O, Sitbon F, Sharma P, Lönneborg A, von Arnold S (2001a) Identification of candidate genes for use in molecular breeding – a case study with the Norway spruce defensin-like gene, *spi 1*. Silvae Genet 50:75–81

Elfstrand M, Fossdal C-G, Sitbon F, Olsson O, Lönneborg A, von Arnold S (2001b) Overexpression of the endogenous peroxidase-like gene *spi* 2 in transgenic Norway spruce plants results in increased total peroxidase activity and reduced growth. Plant Cell Rep 20:596–603

Elfstrand M, Sitbon F, Lapierre C, Bottin A, von Arnold S (2002) Altered lignin structure and resistance to pathogens in *spi 2*-expressing tobacco plants. Planta 214:708–716
Ellis DD (1995) Transformation of conifers. In: Jain SM, Gupta PK, Newton RJ (eds) Somatic embryogenesis in woody plants, vol 1. Kluwer, Dortrecht, pp 227–251
FAO (1999) State of the World's Forests, 1999. Food and Agricultural Organization of the United Nations
Fenning TM, Gershenzon J (2002) Where will the wood come from? Plantation forests and the role of biotechnology. Trend Biotechnol 20:291–296
Finer JJ, Vain P, Jones MW, McMullen MD (1992) Development of the particle inflow gun for DNA delivery to plant cells. Plant Cell Rep 11:323–328
Fossdal CG, Sharma P, Lönneborg A (2001) Isolation of the first putative peroxidase cDNA from a conifer and the local and systemic accumulation of related proteins upon pathogen infection. Plant Mol Biol 47:423–435
Fu X, Duc LT, Fontana S, Bong BB, Tinjuangjun P, Sudhakar D, Twyman RM, Christou P, Kohli A (2000) Linear transgene constructs lacking vector backbone sequences generate low-copy-number transgenic plants with simple integration patterns. Transgen Res 9:11–19
Gao A-G, Hakimi SM, Mittanck CA, Wu Y, Woerner BM, Stark DM, Shah DM, Liang JH, Rommens CMT (2000) Fungal pathogen protection in potato by expression of a plant defensin peptide. Nat Biotechnol 18:1307–1310
Granger CL, Cyr RJ (2001) Characterization of the yeast copper-inducible promoter system in *Arabidopsis thaliana*. Plant Cell Rep 20:227–234
Hadi MZ, McMullen MD, Finer JJ (1996) Transformation of 12 different plasmids into soybean via particle bombardment. Plant Cell Rep 15:500–505
Häggman H, Aronen T, Nikkanen T (1997) Gene transfer by particle bombardment to Norway spruce and Scots pine pollen. Can J For Res 27:928–935
Hak O (1996) Vacuum processing and storage of spruce pollen. Forest Research Report No 136. Ontario Ministry of Natural Resources, Ontario Forest Research Institute, Sault Ste. Marie, Ontario, Canada, 6 pp
Hay I, Lachance D, von Aderkas P, Charest PJ (1994) Transient chimeric gene expression in pollen of five conifer species following microprojectile bombardment. Can J For Res 24:2417–2423
Hjortswang H, Filonova L, Vahala T, von Arnold S (2002) Modified expression of the *Pa18* gene interferes with somatic embryo development. Plant Growth Regul 38:75–82
Högberg K-A, Ekberg I, Norell I, von Arnold S (1998) Integration of somatic embryogenesis in a tree breeding programme – a case study with *Picea abies*. Can J For Res 28:1536–1545
Högberg K-A, Bozhkov PV, Grönroos R, von Arnold S (2001) Critical factors affecting ex vitro performance of somatic embryo plants of *Picea abies*. Scand J For Res 16:295–304
Ingouff M, Farbos I, Lagercrantz U, von Arnold S (2001) *PaHB1* is an evolutionary conserved HD-GL2 homeobox gene expressed in the protoderm during Norway spruce embryo development. Genesis 30:220–230
Jefferson RA (1987) Assaying chimeric genes in plants: the *uidA* gene fusion system. Plant Mol Biol Rep 5:387–405
Klein TM, Wolf ED, Wu R, Sanford JC (1987) High-velocity microprojectiles for delivering nucleic acids into living cells. Nature 327:70–73
Klimaszewska K, Lachance D, Pelletier G, Lelu MA, Seguin A (2001) Regeneration of transgenic *Picea glauca*, *P. mariana*, and *P. abies* after cocultivation of embryogenic tissue with *Agrobacterium tumefaciens*. In Vitro Cell Dev Biol P 37:748–755
Kobayashi A, Koguchi Y, Kanzaki H, Kajiyama SJ, Kawazu K (1994) Production of a new type of bioactive phenolic compound. Biosci Biotechnol Biochem 58:133–134
Kohli A, Leech M, Vain P, Laurie DA, Christou P (1998) Transgene organization in rice engineered through direct DNA transfer supports a two-phase integration mechanism mediated by the establishment of integration hot-spots. Proc Natl Acad Sci USA 95:7203–7208
Lagrimini LM, Vaughn J, Erb WA, Miller SA (1993) Peroxidase overproduction in tomato wound induced polyphenol deposition and disease resistance. Hort Sci 28:218–221

Lam HM, Hsieh MH, Coruzzi G (1998) Reciprocal regulation of distinct asparagine synthetase genes by light and metabolites in *Arabidopsis thaliana*. Plant J 16:345–353
Lanteri S, Belletti P, Lotito S (1993) Storage of pollen of Norway spruce and different pine species. Silvae Genet 42:104–109
Lindgren D, Paule L, Shen XH, Yazdani R, Segerström U, Wallin JE, Lejdebro ML (1995) Can viable pollen carry Scots pine genes over long distances? Grana 34:64–69
Li Y-H, Tremblay FM, Séquin A (1994) Transient transformation of pollen and embryogenic tissues of white spruce (*Picea glauca* (Moench.) Voss) resulting from microprojectile bombardment. Plant Cell Rep 13:661–665
Lomborg B (2001) The skeptical environmentalist. Cambridge Univ Press, Cambridge, UK, 515 pp
Luomajoki A (1993) Climatic adaptation of Norway spruce (*Picea abies* (L.) Karsten) in Finland based on male flowering phenology. Acta For Fenn 242:1–28
Martinussen I (1994) Characterization of Norway spruce (*Picea abies* (L.) Karst.) pollen as a system for transformation and molecular studies. Doctoral Thesis, Department of Plant Physiology and Microbiology, Institute of Biology and Geology, University of Tromsö, Norway
Martinussen I, Junttila O, Twell D (1994) Optimization of transient expression in pollen of Norway spruce (*Picea abies*) by particle acceleration. Physiol Plant 92:412–416
Martinussen I, Bate N, Weterings K, Junttila O, Twell D (1995) Analysis of gene regulation in growing pollen tubes of angiosperm and gymnosperm species using microprojectile bombardment. Physiol Plant 93:445–450
Matzke MM, Aufsatz W, Kanno T, Mette MF, Matzke AJM (2002) Homology-dependent gene silencing and host defense in plants. In: Dunlap JC, Wu C-T (eds) Advances in genetics, vol 46. Homology effects. Academic Press, San Diego, pp 235–275
Mullin TJ, Bertrand S (1998) Environmental release of transgenic trees in Canada – potential benefits and assessment of biosafety. For Chron 74:203–219
Newton RJ, Yibrah HS, Dong N, Clapham DH, von Arnold S (1992) Expression of an abscisic acid responsive promoter in *Picea abies* (L.) Karst. following bombardment from an electric discharge particle accelerator. Plant Cell Rep 11:188–191
Nikkanen T, Aronen T, Häggman H, Venäläinen M (2000) Variation in pollen viability among *Picea abies* genotypes – potential for unequal paternal success. Theor Appl Genet 101:511–518
Oaks A, Hirel B (1985) Nitrogen metabolism in roots. Annu Rev Plant Physiol 36:345–365
Ortloff W (1999) Sustainability issues in Switzerland's forests. New For 18:59–73
Potrykus I (1991) Gene transfer to plants: assessment of published approaches and results. Annu Rev Plant Physiol Plant Mol Biol 42:205–225
Robertson D, Weissinger AK, Ackley R, Glover S, Sederoff RR (1992) Genetic transformation of Norway spruce (*Picea abies* (L.) Karst) using somatic embryo explants by microprojectile bombardment. Plant Mol Biol 19:925–935
Russell JA, Roy MK, Sanford JC (1992) Major improvements in biolistic transformation of suspension-cultured tobacco cells. In Vitro Cell Dev Biol 28P:97–105
Sabala I, Elfstrand M, Farbos I, Clapham D, von Arnold S (2000) Tissue-specific expression of *Pa18*, a putative lipid transfer protein gene, during embryo development in Norway spruce (*Picea abies*). Plant Mol Biol 42:461–478
Sarvas R (1968) Investigations on the flowering and seed crop of *Picea abies*. Comm Inst For Fenn 67(5):1–84
Sevola Y (ed) (1998) Finnish statistical yearbook of forestry. Finnish Forest Research Institute, Gummerus Kirjapaino Oy, Jyväskylä, 344 pp
Sharma P, Lönneborg A (1996) Isolation and characterization of a cDNA encoding a plant defensin-like protein from Norway spruce. Plant Mol Biol 31:707–712
Srivastava V, Ow DW (2001) Single-copy primary transformants of maize obtained through the co-introduction of a recombinase-expressing construct. Plant Mol Biol 46:561–566
Stöger E, Fink C, Pfosser M, Herbele-Bors E (1995) Plant transformation by particle bombardment of embryogenic pollen. Plant Cell Rep 14:273–278
Strauss SH, Rottmann WH, Brunner AM, Sheppard LA (1995) Genetic engineering of reproductive sterility in forest trees. Mol Breed 1:5–26

Strauss SH, Campbell MM, Pryor SN, Coventry P, Burley J (2001) Plantation certification and genetic engineering: FSC's ban on research is counterproductive. J For 99(12):4–7
Sweet GB, Dickson RL, Donaldson BD, Litchwark H (1992) Controlled pollination without isolation – a new approach to the management of radiata pine seed orchards. Silvae Genet 41:95–99
Sweet GB, Dickson RL, Setiawati YGB, Siregar IZ (1993) Liquid pollination in *Pinus*. For Gen Res Inf 21:2–5
Terras FRG, Eggermont K, Kovaleva V, Raikhel NV, Osborn RW, Kester A, Rees SB, Torrekens S, Van Leuven F, Vanderleyden J, Cammue BPA, Broekaert WF (1995) Small cysteine-rich antifungal proteins from radish: their role in host defense. Plant Cell 7:573–585
van der Leede-Plegt L, van de Ven BCE, Schilder M, Franken J, van Tunen AJ (1995) Development of a pollen-mediated transformation method for *Nicotiana glutinosa*. Transgen Res 4:77–86
Van Loon LC, van Strien EA (1999) The families of pathogenesis-related proteins, their activities and comparative analysis of PR-1 proteins. Physiol Mol Plant Pathol 55:85–97
Vischi M, Marchetti S (1997) Strong extracellular nuclease activity displayed by barley (*Hordeum vulgare* L.) uninucleate microspores. Theor Appl Genet 95:185–190
von Arnold S, Bozhkov P, Clapham D, Dyachok J, Filonoa L, Höberg A, Ingouff M, Wiweger M (2002) Propagation of Norway spruce via somatic embryogenesis. Conference proceedings, Oslo, Norway, 2002
Walter C, Grace LJ, Donaldson SS, Moody J, Gemmell JE, van der Maas S, Kvaalen H, Lönneborg A (1999) An efficient biolistic transformation protocol for *Picea abies* embryogenic tissue and regeneration of transgenic plants. Can J For Res 29:1539–1546
Wenck AR, Quinn M, Whetten RW, Pullman G, Sederoff R (1999) High efficiency *Agrobacterium*-mediated transformation of Norway spruce (*Picea abies*) and loblolly pine (*Pinus taeda*). Plant Mol Biol 39:407–416
Woodward S, Stenlid J, Karjalainen R, Hütterman A (1998) *Heterobasidion annosum*: biology, ecology, impact and control. CAB International, Wallingford, UK
Yibrah HS, Manders G, Clapham DH, von Arnold S (1994) Biological factors affecting transient transformation in embryological suspension cultures of *Picea abies*. J Plant Physiol 144: 472–478

9 WHISKERS-Mediated Transformation of Maize

J.F. PETOLINO, M. WELTER, and C. QIHUA CAI

9.1 Introduction

Since the advent of *Agrobacterium* transformation of cereal crop species via the "super binary" system (Hiei et al. 1994), direct DNA delivery has gone somewhat "out of vogue" for transgenic maize production. The simplicity of application, the range of available options relative to target tissue, and most importantly, the quality of the subsequent transgenic events with respect to simple integration make *Agrobacterium* the current transformation method of choice for maize (Ishida et al. 1996). Nonetheless, direct DNA delivery is still a viable option for transgenic maize production (Petolino 2002) as well as for nontransgenic nucleic acid-based gene modification (Zhu et al. 1999).

WHISKERS-mediated gene transfer, by virtue of its simplicity and potential for scale-up, is an attractive means of delivering DNA to plant cells. Transgenic maize plants and progeny have been recovered from WHISKERS transformation of embryogenic suspension cultures (Frame et al. 1994) as well as friable embryogenic callus (Petolino et al. 2000). However, even prior to the establishment of *Agrobacterium* transformation, WHISKERS-mediated DNA delivery was not the most popular method for transgenic maize production. Low frequencies of transgenic colony formation, strict dependence on specific target tissue cultures, and complex transgene integration made the system somewhat unattractive. In spite of these limitations, a few laboratories have persisted with the goal of developing WHISKERS as an efficient, high-throughput means of transgenic maize production (Bullock et al. 2001; Song et al. 2002).

The present chapter focuses on recent developments in WHISKERS-mediated transformation of maize with particular emphasis on scaling-up the process to produce large numbers of transgenic colonies and rapid nucleic acid screening procedures to select for events displaying "low copy" DNA integration.

9.2 Preparation of Puri ed DNA Fragments

Transformation vectors are typically plasmids that contain one or more plant transformation units (PTUs) consisting of the coding region(s) of the gene(s)

Molecular Methods of Plant Analysis, Vol. 23
Genetic Transformation of Plants

to be transferred and the appropriate genetic regulatory elements required for expression in plants. In addition to the PTUs, most vectors used for direct DNA delivery into plant cells contain vector backbone sequences such as a bacterial selectable marker gene (i.e., β-lactamase for ampicillin resistance) to facilitate cloning in *E. coli*. In order to generate antibiotic resistance gene-free transgenics, it is necessary to use isolated PTU-containing DNA fragments that are free of contaminating backbone sequences. Although such fragments can be readily generated via restriction enzyme digestion of carefully designed vectors, effective recovery of isolated PTUs from agarose gels is not practical for large-scale purification. In order to generate the quantities of isolated PTU-containing DNA fragments required for WHISKERS transformation (several milligrams per study), a column-based system was developed.

Vertically mounted Fast Protein Liquid Chromatography (FPLC) columns (Pharmacia Biotech, Uppsala, Sweden) packed with an aqueous size-exclusion resin allows for the separation of large molecules based on differences in relative size. Sephacryl S-1000 SF is a gel filtration resin commercially available in 20% ethanol (Pharmacia Biotech, Uppsala, Sweden). Before use, the resin is passed through a 0.45-µm filter to remove the ethanol, resuspended in "elution buffer" (1 × TE supplemented with 150 mM NaCl) to 70% v/v, and degassed using a vacuum. The packed bed volume is determined based on column size (an XK50/100 column has a bed volume of 1960 ml) and about 150% of this volume is poured into an extension reservoir attached to the column. The XK50/100 column is then attached to an FPLC system and packed at a rate of 15 cm/h. After two bed volumes (~4000 ml), the extension reservoir is removed and a flow adapter is inserted so as to just touch the upper surface of the resin. The column is then packed for an additional two bed volumes and the flow adapter is readjusted to be sure that no excess buffer remains at the top of the column.

Plasmid DNA is digested with appropriate restriction enzymes that can remove the PTU-containing fragment from other backbone sequences such as the antibiotic resistance gene. Following digestion, using the "elution buffer", the DNA is brought to a concentration of 1.0–1.2 mg/ml and heated to 55°C for 10–15 min before loading 4–5 mg onto the packed XK50/100 column. FPLC monitoring of UV wavelengths is set to 260, 280, and 320 nm. DNA fragments are then eluted at the same flow rate as used for column packing (Fig. 9.1). When a given batch of digested DNA is completely eluted, fractions containing the PTU-containing fragments are pooled and recovered via standard ethanol precipitation.

In order to assess the quality of the isolated PTU-containing fragments, restriction digests are run on a gel to make sure that backbone sequences resulting from partial digestion are not contaminating the samples (Fig. 9.2). The purity of the fragments is estimated by PCR amplification of the antibiotic resistance sequence following serial dilution of the template DNA made from the fragments before and after purification. Typically, a 99% or higher purification is achieved with a DNA recovery rate of about 85%.

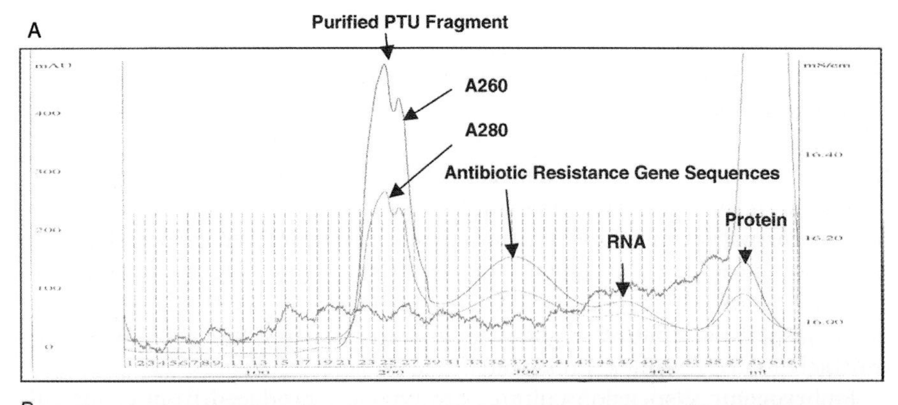

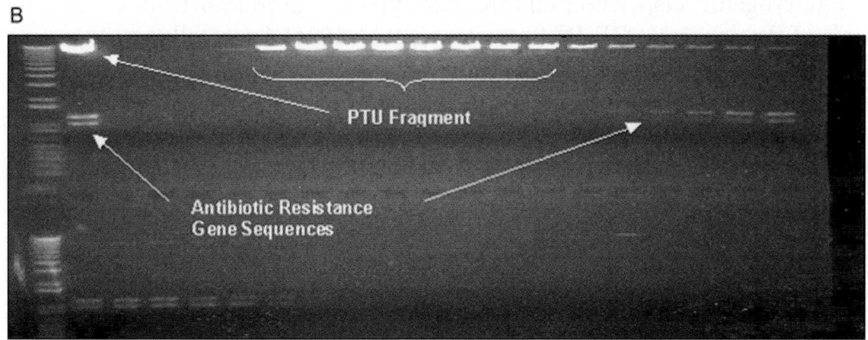

Fig. 9.1A,B. PTU fragment purification via FPLC. **A** Chromatograph of Fsp1-digested DNA showing separation restriction fragments. **B** Gel electrophoresis of eluted DNA

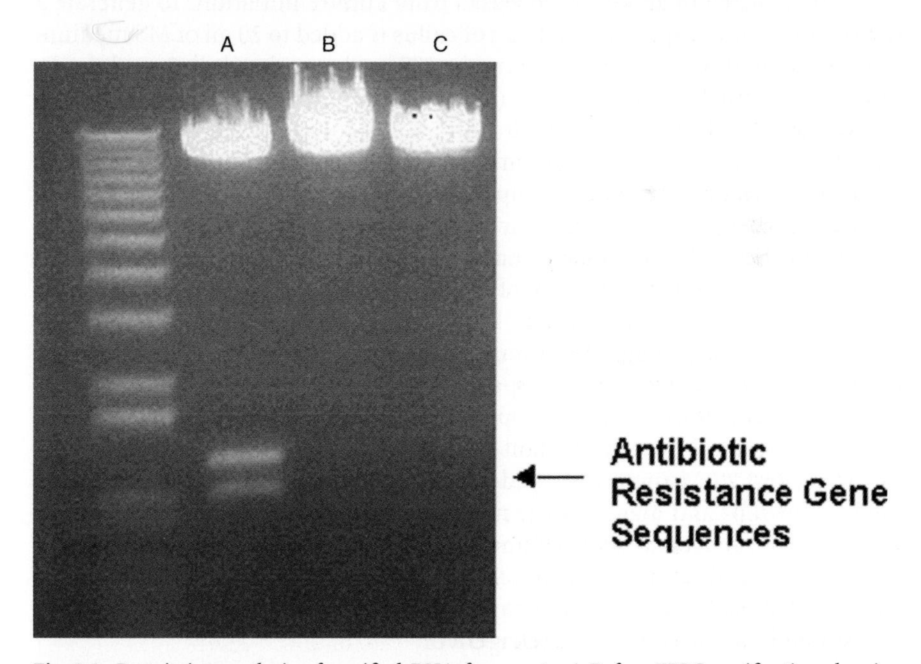

Fig. 9.2. Restriction analysis of purified DNA fragments. *A* Before FPLC purification showing antibiotic resistance gene-containing sequences. *B* and *C* Uncut and Fsp1-digested purified samples, respectively, free of antibiotic resistance gene sequence

9.3 Establishment and Maintenance of Embryogenic Suspension Cultures

Since WHISKERS treatment delivers DNA effectively to surface cell layers only and causes significant physical damage, target tissue cultures amenable to transformation must possess mitotically active surface cells resistant to delivery-induced stress such that transgenic tissue can proliferate during selection and ultimately regenerate fertile plants (Petolino 2002). Maize embryogenic suspension cultures display such characteristics and have been used routinely for WHISKERS transformation.

Embryogenic suspension cultures are typically produced from callus cultures of the genotype "Hi-II" (Armstrong et al. 1991). Such callus cultures are initiated from 1.5–2.0 mm zygotic embryos isolated from ears 10–14 days post-pollination following surface sterilization with 70% ethanol for 2 min and 20% commercial bleach (1% sodium hypochlorite) for 30 min. Callus initiation medium consists of N6 basal salts and vitamins (Chu et al. 1975), 20 g/l sucrose, 25 mM L-proline, 100 mg/l casein hydrolysate, 10 mg/l $AgNO_3$ and 1.0 mg/l 2,4-diclorophenoxy acetic acid (2,4-D) adjusted to pH 5.8 and solidified with 2.0 g/l Gelrite (Aceto Corp., Lake Success, NY). After 8 weeks of 2-week subcultures, the $AgNO_3$ is omitted and the L-proline is reduced to 6 mM for maintenance.

Callus cultures that reach the "early embryogenic" stage of development are most amenable for starting suspension cultures (Welter et al. 1995). This stage is usually reached in about 12–16 weeks from culture initiation. To generate a suspension culture, approximately 3 g of callus is added to 20 ml of MS medium (Murashige and Skoog 1962) containing 100 mg/l myo-inositol, 2 mg/l 2,4-D, 2 mg/l 1-naphthalene acetic acid (NAA), 6 mM L-proline, 200 mg/l casein hydrolysate, 30 g/l sucrose at pH 6.0 with 5% v/v coconut water (Gibco, Grand Island, NY) added at each subculture. The callus is broken up to suspend the tissue in the medium by pipetting up and down several times with a wide-bore 10 ml pipette. Suspension cultures are maintained in 125-ml Erlenmeyer flasks at 28°C in the dark on a rotary shaker at 125 rpm. Subculture is performed every 3.5 days by transferring 3 ml of settled cells and 7 ml of old (conditioned) medium to 20 ml of fresh medium. When suspension cultures reach approximately 32–36 weeks in age (from immature embryo callus initiation), they are ready for use in transformation experiments (Fig. 9.3A).

Suspension cultures can be cryopreserved at liquid nitrogen temperatures. An equal volume of cryoprotectant solution (1.0 M glycerol, 1.0 M dimethylsulfoxide, and 2.0 M sucrose) is added to 2-day post-subculture embryogenic suspension cells and allowed to incubate for 1 h at 4°C on a rotary shaker at 125 rpm. After incubation, the mixture is dispensed in 4.5-ml aliquots into 5.0-ml cryogenic vials and kept on ice. The vials are cooled using a temperature-controlled freezer at a rate of 0.5°C/min until a temperature of –40°C is reached and stored in a liquid nitrogen refrigerator.

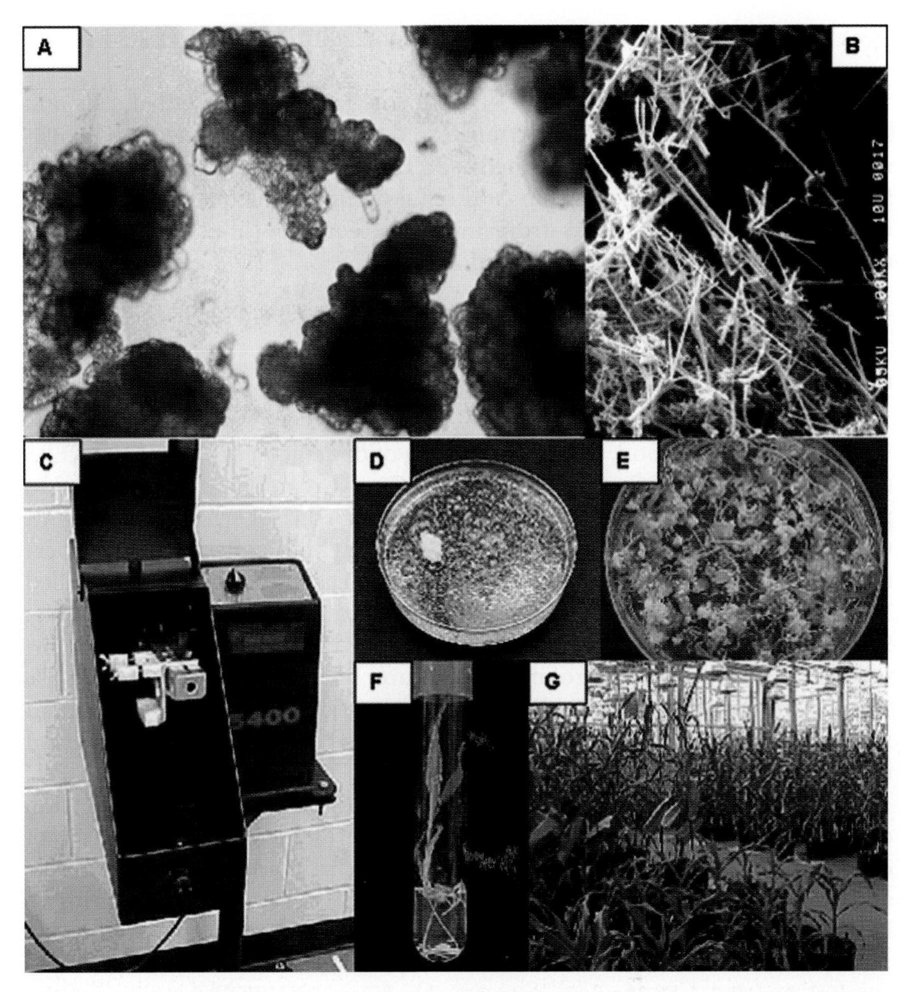

Fig. 9.3A—G. Transgenic maize production via WHISKERS transformation of embryogenic suspension cultures. **A** Embryogenic suspension culture (100×), **B** scanning electron micrograph of silicon carbide WHISKERS, **C** modified Red Devil paint mixer, **D** transgenic event growing on selection plate, **E** shoot regeneration, **F** plantlet on rooting medium, **G** transgenic plants growing to maturity

When required for use, the samples are removed from storage and thawed by plunging the cryogenic vials into a 45°C water bath. Once the samples are completely thawed, the contents of the tubes are poured across the surface of stacked sterile 5.5-cm filter paper discs (Whatman No. 4) in a 60 × 20 mm Petri dish. Excess cryoprotectant is removed by repeated blotting with filter paper. Suspension culture cells on the filter paper are placed on solid N6 medium with 100 mg/l myo-inositol, 2 mg/l 2,4-D, 30 g/l sucrose at pH 6.0 and subcultured weekly for 2–3 weeks after which the suspension cultures are reinitiated as previously described.

Four weeks prior to WHISKERS transformation, 12 ml of settled cells and 28 ml of conditioned medium are transferred to 80 ml of fresh medium in 500-ml Erlenmeyer flasks and maintained at 28°C in the dark at 125 rpm. The useful life of a typical maize suspension culture (assuming fertile plant regeneration is necessary) is 48–56 weeks from immature embryo–callus initiation (not counting time spent in cryopreservation).

9.4 DNA Delivery via WHISKERS

In a chemical fume hood, about 410–420 mg of dry silicon carbide WHISKERS (Silar SC-9, Advanced Composite Materials Corp, Greer, SC) are transferred to preweighed 30-ml polypropylene centrifuge tubes (Fisher, Pittsburgh, PA). Gloves and a respirator are worn while the weighing and transfer are performed and damp paper towels are spread out so as to immobilize any spilled WHISKERS which, when dry, represents a serious respiratory hazard (Fig. 9.3B). The centrifuge tubes containing the WHISKERS are autoclaved on a slow release cycle and stored at room temperature. Immediately before use, a 5% w/v suspension is made by adding sterile "osmotic medium" (see below) and vortexing at maximum speed for 60 s. Suspended WHISKERS is transferred using wide-bore pipette tips.

Four weeks prior to WHISKERS transformation, embryogenic suspension cultures are scaled-up by transferring 12 ml of settled cells and 28 ml of conditioned medium to 80 ml of fresh medium in 500-ml Erlenmeyer flasks and maintained at 28°C in the dark at 125 rpm. Twenty-four hours prior to WHISKERS treatment, 12 ml of settled cell volume and 28 ml of conditioned medium are transferred to 80 ml of N6 liquid medium with 100 mg/l myo-inositol, 2 mg/l 2,4-D, 30 g/l sucrose. On the day of WHISKERS treatment, the cells are given an osmotic pretreatment by drawing off the conditioned medium, replacing it with 72 ml of osmotic medium (same medium with the addition of 0.25 M sorbitol and 0.25 M mannitol), and incubating for 30 min at 125 rpm. Following the osmotic pretreatment, the contents of three flasks are pooled into one sterile 250-ml-IEC centrifuge bottle and allowed to settle for 3–5 min. Approximately 200 ml of osmotic medium is then drawn off leaving approximately 50 ml of cells and medium at the bottom of the centrifuge bottle. The osmotic medium is saved in a clean sterile flask to be reused later on during recovery.

Transformation is carried out by adding 8.1 ml of freshly prepared 5% WHISKERS suspension and 170 μg of DNA. The bottle is immediately transferred to a modified paint mixer (Red Devil Equipment Co., Minneapolis, MN) in which the paint can clamp assembly has been retrofitted (Fig. 9.3C). The bottle is agitated at maximum speed for 10 s after which the cells are returned to a 1-l recovery flask in which the solution is diluted with the reserved osmotic medium along with 125 ml of fresh liquid N6 medium to a final volume of 375 ml. WHISKER-treated cells are allowed to recover for 2 h at 125 rpm at 28°C in the dark.

Following recovery, 15-ml aliquots of the cell suspension culture are evenly dispensed on 125-mm No. 4 filter paper discs (Whatman International Ltd., Maidstone, UK) resting on a two-piece Buchner funnel and liquid medium aspirated through a filter flask. The filter papers with cells are then transferred to 150 × 15-mm Petri dishes containing semi-solid N6 medium with 2 mg/l 2,4-D, 100 mg/l myo-inositol, 30 g/l sucrose, and 2.0 g/l Gelrite at pH 6.0. Each bottle will result in 25 filter plates. Plates are then wrapped with gas-permeable micropore surgical tape (3 M Corporation, St. Paul, MN) and incubated at 28°C in the dark. After 1 week, the filter paper and cells are transferred to fresh N6 medium with the addition of 5.0 mg/l Herbiace (Meiji, Tokyo, Japan). This is repeated for one more week after which the cells are embedded onto semi-solid N6 medium with 2.5 mg/l Herbiace.

For embedding, the tissue on each filter paper is equally divided into five sections and scraped (using a sterile spatula) into a sterile 50-ml centrifuge tube containing 7 ml of N6 liquid medium, 2.5 mg/l Herbicace, and 7% melted SeaPlaque agarose (BMA, Rockland, ME). The tissue is then broken up and resuspended using a 10-ml-wide bore pipette, and evenly dispersed onto the surface of 150 × 15-mm Petri dishes containing semi-solid N6 medium with 5.0 mg/l Herbiace. Once the agarose has solidified, each plate is wrapped in Nescofilm (Azwell Inc., Osaka, Japan) and maintained in the dark at 28°C. Herbiace-resistant callus colonies are observed 3–5 weeks after embedding (Fig. 9.3D). Callus colonies are individually transferred to 60 × 20-mm Petri dishes containing fresh N6 medium with 5 mg/l Herbiace and subcultured every 2 weeks. When sufficient tissue is available (~50 mg fresh weight), the culture is designated as a putative transgenic event and samples are submitted for copy number determination.

9.5 Transgene Copy Number Estimation

Transgenic plants that carry multiple copies of integrated DNA are likely to exhibit transgene silencing (Iyer et al. 2000). Since direct DNA delivery methods, such as WHISKERS, typically result in transgenic events with complex patterns of transgene integration, a reliable method for the rapid determination of copy number would be highly desirable for eliminating multiple copy events early in the transgenic production process. Recent advances in polymerase chain reaction (PCR) instrumentation and fluorescence chemistry have made the precise quantification of specific amplification products possible. These developments have enabled the rapid and efficient estimation of the transgene copy number (Chiang et al. 1996). Quantitative real time PCR (qRT-PCR) technology relies on the ability to progressively monitor fluorescence emitted from specific double-stranded DNA (dsDNA) binding dyes or fluorophore-labeled probes that hybridize with target sequences during the exponential phase of the PCR reaction such that quantification is accomplished. This technology has been adapted for high throughput analysis of transgenic maize callus cultures (Song et al. 2002).

The basic procedure involves establishing a standard curve for transgene copy number using qRT-PCR. This can be accomplished by mixing nontransgenic genomic DNA with plasmid DNA corresponding to 0, 1, 2, 4, 8, and 16 copies of the transgene sequence based on estimated genome size (Fig. 9.4A). The amount of plasmid DNA in pg to make 1000-μl of a one-copy standard with 3 ng/μl of nontransgenic genomic DNA can be calculated as follows: (plasmid size $\times 10^6 \times 3$)/genome size (2 N). After testing the effect of different amounts of template DNA on PCR reactions, it was determined that 6 ng of genomic maize DNA was an appropriate amount in a 20-μl reaction. Assuming that the genome size for maize is 2.5×10^9 bp (Armuganathan and Earle 1991), the amount of plasmid DNA needed to mix with 6 ng of nontransgenic genomic maize DNA to set up the standard curve could be calculated. Once standard curves were established, copy number differences between transgenic events could be estimated.

Genomic DNA extraction from maize callus can be performed using a DNeasy 96 Plant Kit (Qiagen Inc., Vaencia, CA). Using this method, anywhere from 3–50 μg of genomic DNA can be typically obtained from 10–20 mg of freeze-dried or 50–100 mg of fresh maize callus tissue. The fluorescent nucleic acid stain PicoGreen (Molecular Probes, Eugene, OR) is used to quantify dsDNA in solution using a Microplate Fluorescence Reader (BioTek Instruments, Winooski, VT) with excitation and emission wavelengths of 485 and 530 nm, respectively.

PCR primers can be designed using Vector NTI software (InforMax, North Bethesda, MD), synthesized by Integrated DNA Technologies (Coralville, IA), and diluted to 20 μM in sterile distilled water. Although any PCR product under 1000 bp will work for qRT-PCR, products between 300–500 bp are preferred. In addition, the GC content may significantly affect the efficiency of amplification so that it should not exceed 65%.

A SYBR Green I/LightCycler System (Roche Molecular Biochemicals, Indianapolis, IN) is used to perform qRT-PCR. SYBR Green I is a dye, the fluorescence of which is enhanced when it binds to the minor groove of dsDNA. Using a LightCycler, fluorescence intensity can be measured during various cycles of PCR. In this way, the increase in fluorescence is a function of the amount of dsDNA present, which is proportional to the initial concentration of template in the sample. Comparing back to the standard curve allows for copy number estimation.

Genomic DNA is diluted to 3 ng/μl with sterile distilled water before being used as a template for qRT-PCR. All reactions were performed in a 20-μl capillary tube containing 2 μl of genomic DNA (6 ng) and 18 μl of PCR

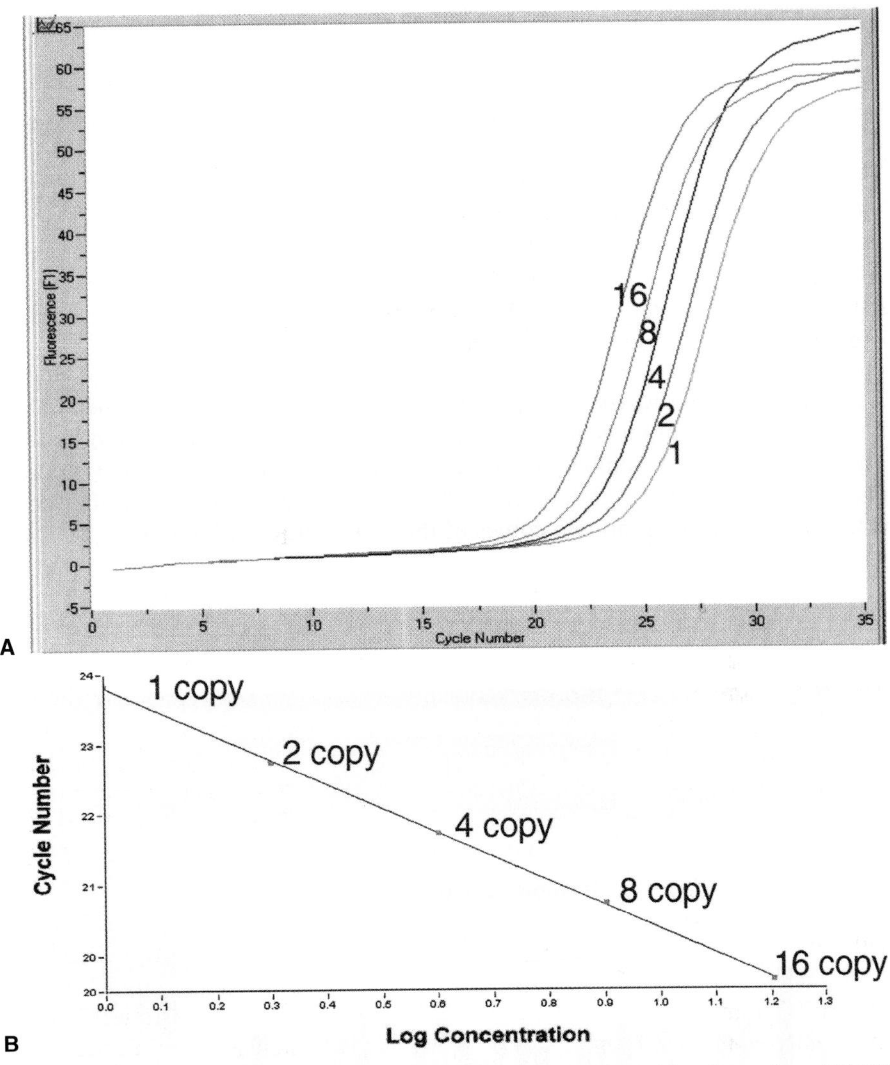

Fig. 9.4A,B. Development of standard curve for copy number estimation using qRT-PCR. **A** Florescence increase over cycles with increasing gene copies. **B** Standard curve

Master-Mixture. The Master-Mixture contained 2 µl PCR buffer from the Fast-Start DNA Master SYBR Green I Kit (Roche Molecular Biochemicals, Indianapolis, IN), 0.5 µl of each primer, 2 µl 25 mM $MgCl_2$, and 13 µl of sterile distilled water. The final concentration of primer and $MgCl_2$ in the 20-µl capillary tube was 0.5 µM and 2.5 mM, respectively. Each tube is then sealed, centrifuged at 700 ×g for 5 s before loading into the rotor of the LightCycler. In each run, two negative controls (water and nontransgenic genomic DNA) are included as are a set of known copy number standards. PCR conditions for amplifying DNA was as follows:

Parameter	Conditions
Activation	95°C for 10 min
Denaturation	95°C for 10 s
Annealing	Primer-dependent, 0–10 s
Elongation time/temperature	Length of product/25, 72°C
Cycle number	25–40
Melting curve	From 65 to 95°C, 0.1°C/s
Cooling	40°C, 30 s, 20°C/s

As the reaction proceeds, a linear regression between the cycle number and the log of the copy number is displayed based on the copy number standards (Fig. 9.4B). The copy number of the tested samples is then calculated automatically based on the standard curve (Fig. 9.5). Callus cultures identified to have integrated more than two copies of the transgene are discarded.

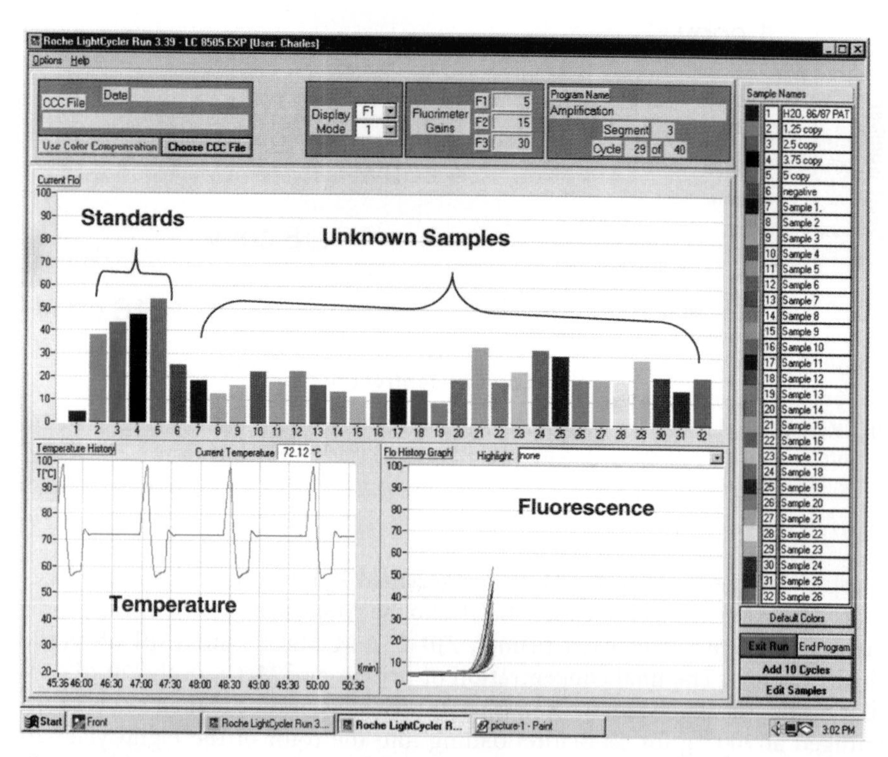

Fig. 9.5. LightCycler readout showing qRT-PCR results for transgene copy number estimation

9.6 Regeneration of Transgenic Plants and Progeny

Transgenic callus cultures identified to have integrated 1–2 transgene copies are regenerated by first transferring to MS (Murashige and Skoog 1962) medium containing 5.0 mg/l benzyl amino purine and 5.0 mg/l Herbiace for 1 week in low light (13 $\mu E\, m^{-2} s^{-1}$) with a 16-h photoperiod followed by 1 week in higher light (40 $\mu E\, m^{-2} s^{-1}$) before transfer to hormone-free MS medium for shoot formation (Fig. 9.3E). Once shoots reach approximately 3–5 cm in length, they are individually transferred to 150 × 25-mm culture tubes containing SH (Schenk and Hildebrandt 1972) medium supplemented with 30 g/l sucrose, 100 mg/l myo-inositol, and 2.5 g/l Gelrite with pH adjusted to 5.8 (Fig. 9.3F) Once the shoots reach the top of the tube, plantlets are transferred to the greenhouse in 10-cm pots containing approximately 0.25 kg of dry Metro Mix 360 (Scotts Co., Maysville, OH), moistened thoroughly, and covered with clear plastic cups for 2–4 days. At the 3–5-leaf stage, plants are transplanted to 5-gallon pots and grown to maturity as described previously (Pareddy et al. 1997). Self- and/or cross-pollinations are performed as necessary to generate transgenic progeny (Fig. 9.3G).

9.7 Conclusions and Future Perspectives

WHISKERS transformation of embryogenic maize suspension cultures has been scaled up into a technology capable of generating large numbers of transgenic isolates capable of regenerating into fertile transgenic plants. Although most of these transgenic events have complex DNA integration, roughly 20% display relatively simple integration (one to two copies) of which a large proportion have been shown to have intact PTUs. Nucleic acid analysis, particularly PCR, has advanced such that methods for rapidly and accurately estimating transgene copy number are now available. These advances, together with the ability to generate and purify large quantities of backbone-free PTU-containing DNA fragments, have made WHISKERS a practical means of transgenic maize production. Future work should include the development of additional in vitro culture systems amenable to WHISKERS transformation to allow for gene transfer directly into a broader spectrum of maize germplasm.

Acknowledgements. The authors would like to thank Dr. Paul Bullock and the Garst Seed Company for providing the modified Red Devil Paint Mixer.

References

Armstrong C, Green C, Phillips R (1991) Development and availability of germplasm with high Type II culture response. Maize Genet Coop News Lett 65:92–93

Armuganathan K, Earle ED (1991) Nuclear DNA content of some important plant species. Plant Mol Biol Rep 9:208–218

Bullock WP, Dilip D, Bagnall S, Cook K, Teronde S, Ritland J, Spielbauer D, Abbaraju R, Christensen J, Heideman N (2001) A high efficiency maize WHISKER transformation system. Plant and Animal Genome IX Conference, San Diego

Chiang PW, Song WJ, Wu KY, Korenberg JR, Fogel EJ, Van Keuren ML, Lashkari D, Kurnit DM (1996) Use of a fluorescent-PCR reaction to detect genomic sequence copy number and transcriptional abundance. Genome Res 6:1013–1026

Chu CC, Wang CC, Sun CS, Hsu C, Yin KC, Chu CY, Bi FY (1975) Establishment of an efficient medium for anther culture of rice through comparative experiments on the nitrogen source. Sci Sin 18:659–668

Frame BR, Drayton PR, Bagnall SV, Lewnau CJ, Bullock WP, Wilson HM, Dunwell JM, Thompson JA, Wang K (1994) Production of fertile transgenic maize plants by silicon carbide whisker-mediated transformation. Plant J 6:941–948

Hiei Y, Ohta S, Kamari T, Kumashiro T (1994) Efficient transformation of rice (*Oryza sativa L.*) mediated by *Agrobacterium* and sequence analysis of the boundaries of the T-DNA. Plant J 6:271–282

Ishida Y, Saito H, Ohta S, Hiei Y, Komari T (1996) High efficiency transformation of maize (*Zea mays* L.) mediated by *Agrobacterium tumefaciens*. Nat Biotech 14:745–750

Iyer LM, Kumpatla SP, Chandrasekharan MB, Hall TC (2000) Transgene silencing in monocots. Plant Mol Biol 43:323–346

Murashige T, Skoog F (1962) A revised medium for rapid growth and bioassays with tobacco tissue cultures. Physiol Plant 15:473–497

Pareddy D, Petolino J, Skokut T, Hopkins N, Miller M, Welter M, Smith K, Clayton D, Pescitelli S, Gould A (1997) Maize transformation via helium blasting. Maydica 42:143–154

Petolino JF (2002) Direct DNA delivery into intact cells and tissues. In: Khachatourians GG, McHughen A, Scorza R, Nip WK, Hui YH (eds) Transgenic plants and crops. Dekker, New York, pp 137–143

Petolino JF, Hopkins NL, Kosegi BD, Skokut M (2000) Whisker-mediated transformation of embryogenic callus of maize. Plant Cell Rep 19:781–786

Schenk RU, Hildebrandt AC (1972) Medium and techniques for induction and growth of monocotyledonous and dicotyledonous plant cell cultures. Can J Bot 50:199–204

Song P, Cai CQ, Skokut M, Kosegi BD, Petolino JF (2002) Quantitative real time PCR as a screening tool for estimating transgene copy number in WHISKERS-derived transgenic maize. Plant Cell Rep 20:948–954

Welter ME, Clayton DS, Miller MA, Petolino JF (1995) Morphotypes of friable embryogenic maize callus. Plant Cell Rep 14:725–729

Zhu T, Peterson DJ, Taglianai L, St Clair G, Baszynski CL, Bowen B (1999) Targeted manipulation of maize genes in vivo using chimeric RNA/DNA oligonucleotides. Proc Natl Acad Sci USA 96:8768–8773

10 Genetic Transformation of Soybean with Biolistics

D. SIMMONDS

10.1 Introduction

Efficient methods have been developed for the introduction of foreign genes into most crop plants. This technology is useful for both improvement of agronomic traits and fundamental research on gene function. Transformation technology is particularly beneficial for a crop such as soybean, *Glycine max*, that has a very narrow germplasm pool as a result of over 3000 years of breeding. Unfortunately, soybean has been one of the most difficult crop plants to transform. The earliest successful soybean transformations used *Agrobacterium tumefaciens* (Hinchee et al. 1988) and biolistics (McCabe et al. 1988) to deliver the DNA, but the efficiency was very low and the methods were difficult to reproduce. Following this success, various gene delivery methods and target tissues were examined, but efficiency remained low for most approaches (Dinkins et al. 2002). These include methods such as electroporation of protoplasts (Lin et al. 1987; Dhir et al. 1991) and intact nodal tissue (Chowrira et al. 1995, 1996), and sonication-assisted *Agrobacterium*-mediated DNA delivery (Trick and Finer 1998). These methods hold promise, providing plants can be regenerated from protoplasts and transgenic plant recovery is increased. The two most reliable methods for soybean, *A. tumefaciens* and biolistic DNA delivery, are compared and updated in a recent review (Dinkins et al. 2002). The focus of the present review is on biolistic delivery methods, the factors that are critical for improved transformation efficiency, and the remaining challenges. An efficient transformation protocol is included.

To generate transgenic plants, target cells possessing two attributes must be identified; they must be competent for stable integration of foreign DNA and also have the capacity to regenerate to fully fertile plants. Plants that are readily transformable, e.g., tobacco, petunia, *Brassica napus* (Horsch et al. 1985; Moloney et al. 1989), are generally highly regenerable from callus. Following transformation of leaf discs or cotyledon petioles from these species, a short transgenic callus phase precedes organogenesis. Such a system has not been identified for soybean; to date, transgenic soybean plants have been regenerated only from meristematic tissues such as embryonic axes (McCabe et al. 1988), cotyledonary nodes (Hinchee et al. 1988; Di et al. 1996; Zhang et al. 1999; Donaldson and Simmonds 2000; Donaldson et al. 2001) and proliferative embryogenic cultures (Finer and McMullen 1991; Sato et al. 1993; Parrott et al. 1994; Stewart et al. 1996; Maughan et al. 1999; Simmonds and Donaldson 2000; Dinkins et al. 2001; Reddy et al. 2001).

Molecular Methods of Plant Analysis, Vol. 23
Genetic Transformation of Plants

Competent soybean target tissues for particle bombardment have been the meristematic cells of embryonic axes and somatic embryogenic cultures. The regenerable cells of embryonic axes of soybean are 3–5 cell layers below the epidermis and to date have been accessed only by particles from the electric discharge particle device, ACCELL technology (McCabe and Christou 1993). The major advantages of embryonic axes are genotype independence and rapid recovery of plants; however, the recovery of transgene chimeras has been the drawback of this target tissue (Christou et al. 1989; Christou 1990). This system has been described in detail (Christou et al. 1990; Christou 1994) and will not be considered here. The focus here, is on proliferative somatic embryogenesis because these embryos arise from single or a few cells that are close to the surface of young globular and older embryo clusters, and, therefore, provide accessible targets to particles discharged from biolistic devices (e.g., Sanford et al. 1991; Finer et al. 1992).

Skillful application of plant tissue culture and plant regeneration procedures are essential for genetic modification of somatic embryogenic cultures. The major problems encountered are low transformation efficiency and the generation of sterile plants. The cause of these difficulties, in most cases, originates with the cell cultures, e.g., genotypes recalcitrant to tissue culture, slow embryo proliferation, abnormal embryo maturation, germination and plant development, gene delivery at incorrect culture growth phase, etc. These topics, reviewed here, have been the subject of numerous studies and have led to a greater understanding of the critical factors affecting transformation efficiency.

10.2 Tissue Culture and Plant Regeneration

10.2.1 Genotype Speci city

Initiation of embryogenesis and continued embryo proliferation is genotype-dependent (Baily et al. 1993a; Tian et al. 1994; Simmonds and Donaldson 2000; Meurer et al. 2001) and has been highly efficient in relatively few genotypes, e.g., cv. Jack (Meurer et al. 2001). Embryogenesis initiation and continued proliferation are independent genotype traits; some genotypes produce abundant primary embryos, but do not make secondary embryos that continue to proliferate, e.g., Cannatto (Simmonds and Donaldson 2000); other genotypes produce few primary embryos, but make excellent prolific cultures, e.g., X5 (X2650-7-2-3, AAFC soybean breeding line; Simmonds and Donaldson 2000) and PI417138 (Bailey et al. 1993a). A frequently reported dysfunction of somatic embryogenesis is recovery of morphologically aberrant plants with partial to total plant sterility (Widholm 1995; Finer et al. 1996; Hadi et al. 1996; Liu et al. 1996). Chromosomal abnormalities can occur in cells that have been cultured for less than 5 months and plants recovered from such cultures are generally abnormal and sterile; use of these genotypes can be avoided as susceptibility to chromosomal instability is genotype-dependent (Singh et al. 1998). Unfortunately, genotypes that show no chromosome damage, even after 2–3 years in culture, can also produce sterile plants (Singh et al. 1998). Genetic mutations and minor chromosomal variations that are not detectable cytologically have also been attributed to high 2,4-dichlorophenoxyactetic acid (2,4-D) concentrations in cultures (Bayliss 1980). For this reason, it is advisable to use cultures that are less than one year old and to minimize 2,4-D concentrations. However, age can be defined in terms of time (months in culture) or in numbers of generations. It is, therefore, possible to maintain cultures for prolonged periods of time in a "youthful state" by extending the culture generation time (see below).

10.2.2 Initiation and Repetitive Proliferation of Somatic Embryogenic Cultures

Previous studies have documented somatic embryogenesis and plant regeneration from a variety of soybean genotypes (Christianson et al. 1983; Lippman and Lippman 1984; Lazerri et al. 1985; Ranch et al. 1985; Barwale et al. 1986; Komatsuda and Ohyama 1988; Parrott et al. 1988), but these systems were not transformation-competent. The breakthrough for soybean transformation came with the reports showing that somatic embryogenesis could be maintained as repetitive proliferative cultures and the proliferative phase could be arrested at any time to promote embryo development and plant recovery (Finer 1988; Finer and Nagasawa 1988). In this approach, embryogenesis is initiated from cotyledons, dissected from 3–5 mm seed, and cultured at 28°C, with the abaxial side in contact with solid modified MS medium [MS salts (Murashige and Skoog 1962) and B5 vitamins (Gamborg et al. 1968), 6% sucrose, 0.8% agar, pH 5.7] as previously described (Lazzeri et al. 1985). High 2,4-D concentration (40 mg/l) prevented embryo development past the heart stage, but permitted proliferation of secondary globular embryos (Finer 1988). Repetitive proliferation of the initiated embryogenic tissue could be maintained on the same medium, or on FN, a modified MS liquid medium with 2,4-D levels reduced to 5–10 mg/l and modified nitrogen content that included the amino acids glutamine or asparagine, reduced inorganic nitrogen and a lower ammonium/nitrate ratio (Finer and Nagasawa 1988). Based on earlier work showing that wounding and lower sugar concentrations increased embryogenesis frequency (Lazzeri et al. 1987), the induction conditions were examined to optimize physical conditions such as explant orientation, pH, solidifying agent and wounding (Santarem et al. 1997). Wounding of cotyledons, 0.2% gelrite and pH 7.0 resulted in the highest embryo induction with the genotypes tested, however, the optimal pH may be genotype-dependent (Lazzeri et al. 1987; Simmonds, unpubl. observ.). Furthermore, similar to the previous report on embryogenesis induction (Lazzeri et al. 1987), Samoylov et al. (1998a) found that lower sugar concentrations enhanced somatic embryo growth in repetitive proliferative cultures, and in combination with reduced nitrogen levels, greatly improved the FN medium which they renamed FN Lite medium. It is important to point out that these improvements have been carried out on a limited set of soybean genotypes and that optimal conditions for tissue culture should be assessed for each different genotype. Updates on media formulation and other improvements to the soybean tissue culture protocols are posted on the web site of the Soybean Tissue Culture and Genetic Engineering Center (http://mars.cropsoil.uga.edu/homesoybean/).

Improvements in the media formulations for proliferative cultures increased the rate of growth, so that at temperatures of 26–28°C, weekly subculturing was required. This protocol accelerated the generation time. Slower growth rates were achieved by maintaining proliferative cultures on solid media with 2,4-D levels increased from 5–10 to 20 mg/l which provides the advantage of simplified tissue culture manipulations, but has the disadvantage of the higher 2,4-D concentrations and the potential risk of genetic aberrations. The slower growth rate of cultures on high 2,4-D solid media impacts transformation by extending the time required for clone selection (6–8 months) and consequently, the recovery of infertile and/or abnormal plants at high frequencies (Dinkins et al. 2002). An alternative approach (developed at the author's laboratory) that slows down the generation time and results in good recovery of normal fertile plants (see Table 10.1), uses lower

Table 10.1. Relationship of age of different embryogenic cultures at time of bombardment and their transformability and recovery of fertile plants

Culture	Efficiency factor	Culture age at bombardment (months)								
		1–2	3	4–4.5	7–8.5	10	12.5	14	19	24
X5[a]-A7	T[c]				7–12		8–30			
	F[d]				60–100		90–100			
X5-18/15	T		0–9					9–24		8–11
	F		100					100		50–83
X5-13/15	T					2–40			1–12	
	F					100			80–100	
X5-03	T			11–26		15–25				
	F			88–100		0–100				
We[b]-01a	T	0–26		30–37						
	F	100		50–100						
We-02b	T	40–68								
	F	83–100								

[a] X5, AAFC breeding line X2650-7-2-3 (~MG00).
[b] We, OAC Westag 97 (MG2).
[c] T, Transformability of a line: number of transformation events recovered from 50–75 mg of bombarded tissue.
[d] F, Fertility of a line: percent of embryogenic transformation events that produced fertile plants.

2,4-D levels for induction of embryogenesis (20 mg/l instead of 40 mg/l) and low levels for maintenance of proliferative cultures (10 mg/l), and conducts all procedures at the lower temperature of 20°C, including induction of embryogenesis, repetitive embryo proliferation, selection of transgenic events, histodifferentiation, embryo maturation, desiccation and germination (see Sect. 10.5). Reducing temperature from 27 to 23°C has been investigated and has shown improvements in another laboratory (Buenrostro-Nava et al. 1999).

10.2.3 Embryo Histodifferentiation and Maturation

Repetitive proliferation of globular embryos is arrested by removal of 2,4-D. Finer and McMullen (1991) cultured globular embryo clusters on auxin-free media containing maltose (MS salts, B5 vitamins, 6% maltose 0.2% gelrite, pH 5.7) to facilitate embryo histodifferentiation. For the first 4 weeks, activated charcoal can be added to the medium to remove residual auxins. To complete maturation, embryos were separated from the clusters and cultured individually for another 3–4 weeks on the same medium with no charcoal (Bailey et al. 1993a). The incubation temperature varied from 23°C (Finer and McMullen 1991) to 26–28°C (Bailey et al. 1993a). This lengthy procedure was reduced to 3 weeks by carrying out histodifferentiation in FN Lite liquid medium with 3% sucrose at 26°C (Samoylov et al. 1998b). The drawback of this rapid method was a reduction in embryo germination, but this was corrected by increasing the osmoticum of the medium with 3% sorbitol (Walker and Parrott 2001). In a situation where only one or a very few embryos are available, transgenic events may be lost, so the longer, but safer route of plant recovery may be to use the "shoot-bud" culture approach whereby many plants can be regenerated from one somatic embryo (Wright et al. 1991).

10.2.4 Germination, Conversion and Plant Fertility

Germination of mature embryos has been greatly enhanced by the use of partial desiccation (Hammatt and Davey 1987; Parrott et al. 1988). Partial desiccation of embryos has been achieved by placing mature embryos in sealed Petri dishes from 2–3 days at 23°C to 4 weeks at 27 ± 2°C (Hammatt and Davey 1987; Parrott et al. 1988; Finer and McMullen 1991); and germination was further improved at higher relative humidity (RH) by adding a small piece of solid medium to the Petri dishes containing the embryos, but not in contact with the embryos (Trick et al. 1997). Optimal desiccation appears to be genotype-dependent as an RH of 85% produced the best rates of germination for some lines (Bailey et al. 1993b) while 100% RH was ideal for short-season varieties (Simmonds, unpubl.). Following desiccation, embryos have been germinated on hormone-free MS-based medium (MS salts, B5 Vitamins, 0.2% gelrite, pH 5.7–5.8) with 3% sucrose (Finer and McMullen 1991), 1.5% sucrose (Bailey et al. 1993b), or 3% sucrose followed by subculture to 1.5% sucrose (Trick et al. 1997) or 3% sucrose followed by 1% sucrose in B5 media (Dinkins et al. 2002). Embryo quality prior to desiccation determines whether normal germination occurs.

After root and shoot emergence, the plantlets are moved to soil and gradually acclimated to the environment. This procedure (embryo desiccation, germination, planting to soil) takes 6–8 weeks (Trick et al. 1997; Dinkins et al. 2002). With good quality embryos, the procedure can be simplified and the time shortened to 3–7 days for desiccation (100% RH) and 1–2 weeks for germination on B5 with 2% sucrose. Leaving the germinated plantlets in culture media for a prolonged period of time can be detrimental (Simmonds, unpubl. res.); as soon as a small root and shoot emerge, the plantlets can be moved directly to soil (see Sect. 10.5).

10.3 Transformation

10.3.1 Gene Delivery

Delivery of DNA by acceleration of DNA-coated particles into cells was first reported in 1987 (Klein et al. 1987; Sanford et al. 1987). The early gene guns used gunpowder for particle propulsion, but pressurized helium has now replaced the gunpowder in most of the devices. In addition to the commercial helium gun, Biolistic PDS-1000/He Particle Delivery System (Bio-Rad Laboratories, Richmond, CA, USA; Sanford et al. 1991), other gun designs have been developed, including the ACCELL technology (electric discharge particle acceleration; McCabe and Christou 1993), the particle inflow gun (Finer et al. 1992), the microtargetting device (Sautter et al. 1991), the airgun (Oard et al. 1990) and the handheld Helios gene gun (Sundaram et al. 1996). Whichever gene gun is used, the delivery system must be optimized such that maximum gene delivery is achieved with minimum tissue damage. The variables that affect gene delivery by particle bombardment include the microprojectile penetrating ability (size, density, shape), its DNA-carrying ability (DNA quantity, quality and precipitation protocol, and particle size, shape and surface characteristics), its velocity (acceleration force, distance to target, vacuum in the chamber), and the particle number per tissue area (distance to target, number loaded). The reader is referred to a recent review in which these factors are considered in detail (Klein and Jones 1999). The majority of publications report parameters for maximal transient expression, but the conditions for optimal transient expression and stable transformation may differ considerably (Klein and Jones 1999). Each plant system requires empirical optimization for maximum stable transformation.

10.3.2 Target Tissue Optimization

Very little work has been done to maximize tissue competence of soybean proliferative embryogenesis for stable transformation. While it has been reported that stable transformation is generally very low in cultures less than 6 months old (Stewart et al. 1996; Hazel et al. 1998) and high in older cultures (Hadi et al. 1996), our data show that some lines are highly transformable between 1 and 2 months in culture (Table 10.1). However, very little information is available on the impact of subculture frequency, culture growth dynamics, cell cycle stages and culture morphology on transformation efficiency. Bombardment at different stages of the cell cycle in tobacco indicated that maximal transient expression was obtained during mitosis or G2 (Iida et al. 1991). Similarly, transient expression in soybean cells was highest when the mitotic index peaked, 2–6 days after subculture (Hazel et al. 1998). These investigators also characterized other aspects of soybean cell cultures that correlated with transformation amenability. Histological studies revealed cytoplasm-rich cells in the outermost cell layers of embryo clusters in highly transformable lines and vacuolated cells in less transformable lines. Actively dividing cells are generally cytoplasm-rich and the most transformable cultures had the highest mitotic index. As the different culture phenotypes were generated from the same genotype, they demonstrated that traits affecting transformability of a cell line may be acquired during culture.

Cultures are generally pre-treated in media containing sugar alcohols to increase osmotic pressure in order to enhance cell survival after particle bombardment. Both transient and stable transformation of maize was increased by pre-treatment with sorbitol and mannitol (Vain et al. 1993). This procedure most likely plasmolyzes the target cells and minimizes cytoplasm leakage after particle penetration. Simple air drying of soybean cultures has been used to effect a similar state (Finer and McMullen 1991; Santarem and Finer 1999).

10.3.3 Selection

The selectable resistance gene, *hph*, hygromycin phosphotransferase (Gritz and Davies 1983), has provided the most successful method of selection of transgenic cells in soybean embryogenic cultures. Concentration of hygromycin, method and time of application following tissue recovery may vary depending on the genotype and culturing protocol in use. Hygromycin selection in liquid cultures was initiated 1–4 weeks after transformation at a concentration of 50 mg/l (Finer and McMullen 1991; Parrott et al. 1994; Hadi et al. 1996) or 25–30 mg/l (Cho et al. 1995; Simmonds and Donaldson 2000) and transgenic clones were recovered after 4–9 weeks. Transgenic clones were also recovered by initiating selection 1 day after bombardment on 25 mg/l hygromycin (Stewart et al. 1996; Maughan et al. 1999). Stepwise selection is generally used for solid cultures. Within 1–3 weeks after bombardment, cultures are placed on 9–20 mg/l hygromycin for 2–4 weeks, followed by 18–35 mg/l hygromycin for 2–8 months, until green embryogenic colonies are observed (Santarem and Finer 1999; Reddy et al. 2001; Dinkins et al. 2002). The long period of selection on solid medium increases the probability of recovering abnormal/sterile lines (Sing et al. 1998; Dinkins et al. 2002).

Positive selection systems such as the phosphomannose isomerase gene (*pmi*) with mannose as the carbohydrate source have been successful in recovering transgenic maize plants (Negrotto et al. 2000) and may be useful for soybean. Such an approach would eliminate the need for antibiotic resistance genes, one of the perceived concerns in genetically modified food crops. Other approaches that avoid antibiotic resistance genes include GFP selection (Ponappa et al. 1999) and nondestructive GUS selection (Simmonds and Donaldson 2000).

10.3.4 Transgenic Plant Recovery

Regeneration is carried out according to the procedures outlined above. A major problem has been the recovery of transgenic events that produce abnormal plants with partial or complete sterility (Parrott et al. 1994; Finer et al. 1996; Hazel et al. 1998). As noted above, plant sterility may be caused by chromosomal aberrations which is genotype-specific (Singh et al. 1998) and thus can be avoided, or by culture age (Hadi et al. 1996) which may be affected by time on 2,4-D and concentration of 2,4-D. The lengthy selection schedules on solid media with 20 mg/l 2,4-D may exacerbate this problem. We have recovered fertile plants at good frequencies even after transforming 2-year-old cultures (Table 10.1). This may be due to the specific genotypes or because the cultures are maintained in a "youthful state" for a prolonged period of time by maintaining cultures at the lower temperature of 20°C and a lower level of 2,4-D (10 mg/l), thus lengthening the duration of the cell cycle.

High transgene copy number, multiple integrations, rearrangements, co-suppression and gene silencing (e.g., Finer and McMullen 1991; Hadi et al. 1996; Stewart et al. 1996; Maughan et al. 1999; Dinkins et al. 2001, 2002) are major challenges to overcome in order to optimize transformation of soybean and other crop plants (Flavell 1994; Register et al. 1994; Kononov et al. 1997). Other methods of gene delivery produce similar transgene anomalies. *Agrobacterium*-mediated transformation can also result in multiple gene integrations (Donaldson and Simmonds 2000), but extremely high integration and copy numbers are generally detected only with direct DNA transfer. The number of transgene integration sites may be reduced by simply lowering DNA concentration for bombardment (Table 10.2). Because both the vector backbone and the transgene contribute to complex integration patterns, an advantage of biolistics is that the vector backbone can be eliminated from the

Table 10.2. Effect of bombardment of different quantities of DNA on gene integration, expression and plant fertility. Gene expression was determined using the OxO histological test. (Donaldson et al. 2001)

DNA per shot (ng)[a]	No. of clones producing embryos	No. of E[b] clones with band numbers[c]					Clones lost at embryo maturation	Clones producing plants (T_0) OxO+/total	OxO+ clones lost to plant sterility	OxO+ clones fertile (T_1)
		0	1–2	3–5	6–10	>10				
83	7	2	0	0	1	4	2	3/5	2	1
8.3	8	1	2	3	2	0	1	5/7	0	5
0.83	6	5	1	0	0	0	0	1/6	0	1

[a] DNA fragments [promoter (double 35S), coding regions (OxO), 35S terminator] without the plasmid backbone were used for bombardment.
[b] E, Embryogenic.
[c] Gene integration numbers estimated from number of bands on southern blots.

delivery cassette (Table 10.2). A hybrid of *Agrobacterium* and direct gene delivery, "agrolistics", may help to reduce multiple integrations and copy numbers. Hansen and Chilton (1996) co-bombarded a plasmid carrying a gene flanked by T-DNA border sequences and a second plasmid carrying *Agrobacterium* virulence genes. They showed evidence that close to 20% of the transformants contained gene integrations that were spliced as would be expected in *Agrobacterium*-mediated transformation; the transgenic events also contained low copy numbers. This is a very promising modification to biolistic delivery of DNA.

10.4 Conclusions

The efficiency of soybean transformation has been improved sufficiently to enable routine use for crop improvement and gene expression studies. The major remaining challenge is to develop methods to eliminate gene co-suppression and silencing. Further work on target tissue optimization may also improve transformation efficiency. In spite of the recalcitrance of soybean transformation, important agronomic traits have been introduced, including genes conferring disease resistance (Donaldson et al. 2001; Reddy et al. 2001; Cober et al. 2002), herbicide resistance (Delannay et al. 1995; Padgette et al. 1995), insect resistance (Parrott et al. 1994; Stewart et al. 1996; Walker et al. 2000) and seed quality improvement (Maughan et al. 1999; Dinkins et al. 2001). This is clear evidence of the tremendous potential of this technology for improvement of this very important crop plant.

10.5 Protocol

10.5.1 Induction and Maintenance of Proliferative Embryogenic Cultures

Immature pods, containing embryos 3–5 mm long, are harvested from host plants grown at 26–28/24°C (day/night), 15-h photoperiod provided by cool white fluorescent tubes and incandescent bulbs at a light intensity of 300–400 μmol m^{-2} s^{-1}. Pods are sterilized for 30 s in 70% ethanol followed by 15 min in 2.5% calcium hypochlorite or 1% sodium hypochlorite [with 1–2 drops of Tween 20 (Sigma, Oakville, ON, Canada) in 100 ml] and three rinses in sterile water. The embryonic axis is excised and explants are cultured with the abaxial surface in contact with the induction medium [MS salts, B5 vitamins, 1.25–3.5% glucose (concentration varies with genotype), 20 mg/l 2,4-D, pH 5.7], ten explants per 20 × 60-mm Petri plate. The explants, maintained at 20°C, at a 20-h photoperiod under cool white fluorescent lights at 35–75 μmol m^{-2} s^{-1}, are subcultured four times, at 2-week intervals. The first appearance of embryogenic clusters showing proliferating globular structures on immature secondary embryos can be observed after 3–8 weeks of culture; this time period is genotype-dependent.

Embryogenic clusters are transferred to 125-ml Erlenmeyer flasks containing 30 ml of embryo proliferation medium, FN Lite [(Samoylov et al. 1998a) containing 5 mM asparagine, 1–2.4% sucrose (concentration is genotype-dependent), 10 mg/l 2,4-D, pH 5.0] and cultured as above at 35–60 μmol m^{-2} s^{-1} on a rotary shaker at 125 rpm. Embryogenic tissue (30–60 mg) is carefully selected, using an inverted microscope, for subculture every 4–5 weeks.

10.5.2 Transformation

The earliest and latest culture age for optimal frequency of transformation and fertile plant recovery ranges between 1 and 6 months and 1–2 years, or longer, in liquid culture, respectively, for cvs. Westag 97 and X5 (X2650-7-2-3, an old AAFC breeding line: Table 10.1). Cultures are bombarded 3 days after subculture. Embryogenic clusters from four flasks are used for one experiment. All the embryogenic clusters from one flask are blotted on sterile Whatman filter paper to remove the liquid medium, placed inside a 10 × 30-mm Petri dish on a 2 × 2-cm^2 tissue holder (PeCap, 1005 μm pore size, B and SH Thompson and Co. Ltd. Scarborough, ON, Canada) and covered with a second tissue holder that is then gently pressed down to hold the clusters in place. Immediately before the first bombardment, the tissue is air dried in the laminar air flow hood with the Petri dish cover off for no longer than 5 min. The tissue, sandwiched between the two holders, is turned over, dried as before, bombarded on the second side and returned to the same culture flask.

The bombardment conditions used for the Biolistic PDS-1000/He Particle Delivery System are as follows: 29 in Hg (737 mm Hg) chamber vacuum pressure, 13 mm distance between rupture disc (Bio-Rad Laboratories Ltd., Mississauga, ON, Canada) and macrocarrier; the first bombardment uses 900 psi rupture discs and a microcarrier flight distance of 8.2 cm, and the second bombardment uses 1100 psi rupture discs and 11.4-cm microcarrier flight distance.

DNA precipitation onto 1.0-μm-diameter gold particles is carried out as follows: 2.5 μl of 100 ng/μl test DNA and 2.5 μl of 100 ng/μl selectable marker DNA (*hyg*) are added to 3 mg gold suspended in 50 ul sterile dH_2O in 0.5 ml microfuge tube and vortexed for 10 s; 50 μl of 2.5 M $CaCl_2$ is added, vortexed for 5 s, followed by the addition of 20 μl of 0.1 M spermidine which is also vortexed for 5 s. The mixture is kept suspended by gently flicking the tube, as necessary, for 5 min. The gold is then allowed to settle to the bottom of the microfuge tube (5–10 min) and the supernatant fluid is removed. The gold/DNA is resuspended in 200 μl of 100% ethanol, allowed to settle and the supernatant fluid is removed. The ethanol wash is repeated and the supernatant fluid is removed. The sediment is resuspended in 120 μl of 100% ethanol and aliquots of 8 μl are added to each macrocarrier. The gold is resuspended before each aliquot is removed. The macrocarriers are placed under vacuum to ensure complete evaporation of ethanol (~5 min). A fine lawn of gold/DNA is present on the macrocarriers (the presence of aggregates usually signifies impure DNA that requires further purification). The microcarrier/macrocarrier preparations are used within 40 min.

10.5.3 Selection

The bombarded tissue is cultured on FN Lite medium for 12 days prior to subculture to selection medium (FN Lite, as above, containing 55 mg/l hygromycin added to autoclaved media). The tissue is subcultured 5 days later and weekly for the following 9 weeks. Green colonies generally begin to appear between week 3 and 4 on selection medium. Each colony (putative transgenic event) is transferred to a well containing 1 ml of selection media in a 24-well multi-well plate that is maintained on a flask shaker as above. Colonies continue to appear in the selection medium and are transferred to multi-well plates weekly for the following 4–5 weeks or until 100 individual events have been selected. The media in multi-well dishes is replaced with fresh media every 2 weeks until the colonies are approx. 2–4 mm in diameter and have proliferative embryos, at which time they are transferred to 125 ml Erlenmeyer flasks containing 30 ml of selection medium. Generally, approx. 50% of the selected putative transgenic colonies are embryogenic. The embryogenic colonies are proliferated until sufficient material is available for embryo maturation.

10.5.4 Plant Regeneration

Maturation of embryos is carried out, without selection, at environmental conditions described for embryo induction. Embryogenic clusters are cultured on

20 × 60-mm Petri dishes containing maturation medium similar to that described by Finer and McMullen (1991) and Bailey et al. (1993a; MS salts, B5 vitamins, 6% maltose, 0.2% gelrite gellan gum (Sigma), 750 mg/l $MgCl_2$, pH 5.7) with 0.5% activated charcoal for 5–7 days and without activated charcoal for the following 3 weeks. Embryos (10–15 per event) with apical meristems are selected under a dissection microscope and cultured on a similar medium containing 0.6% phytagar (Gibco, Burlington, ON, Canada) as the solidifying agent, without the additional $MgCl_2$, for another 2–3 weeks or until the embryos, initially green, become pale yellow in colour or partly yellow. Alternatively, embryos can undergo maturation in liquid medium without selection as described by Samoylov et al. (1998b) and Walker and Parrott (2001).

Mature embryos are desiccated by transferring embryos from each event to empty 15 × 60-mm Petri dish bottoms that are placed inside Magenta boxes (Sigma) containing several layers of sterile Whatman filter paper flooded with sterile water, for 100% RH. The Magenta boxes are covered and maintained in darkness at 20°C for 5–7 days. The embryos are germinated on solid B5 medium containing 2% sucrose, 0.2% gelrite and 0.075% $MgCl_2$ in 25 × 100-mm Petri plates (ten embryos or fewer per plate), in a chamber at 20°C, 20-h photoperiod under cool white fluorescent lights at 35–75 $\mu mol\,m^{-2}\,s^{-1}$. Within a few days, the roots emerge followed by shoots. Germinated embryos with unifoliate or trifoliate leaves are planted in artificial soil (Sunshine Mix No. 3, SunGro Horticulture Inc., Bellevue, WA, USA), one plantlet per cell of a 72-cell greenhouse flat (100-K, Kord products, Bramalea, ON, Canada) that is covered with a transparent plastic lid to maintain high humidity. The flats are placed in a controlled growth cabinet at 26/24°C (day/night), 18 h photoperiod provided by cool white fluorescent lights and incandescent bulbs at a light intensity of 150 $\mu mol\,m^{-2}\,s^{-1}$. At the 2–3 trifoliate stage (2–3 weeks) the plantlets have strong roots and are transplanted to 12.5-cm fibre pots containing a 3 : 1 : 1 : 1 mix of ASB Original Grower Mix (a peat-based mix from Greenworld, ON, Canada) : soil : sand : perlite and grown at 18-h photoperiod provided by cool white fluorescent lights and incandescent bulbs at a light intensity of 300–400 $\mu mol\,m^{-2}\,s^{-1}$. The photoperiod is reduced to 13 h for flower induction when the plants are at the V 5–6 stage and approx. 20–25 cm high (approx. 2 weeks after potting). Growth at these two photoperiods is important for production of fertile plants because plants can be sterile if moved to the short photoperiod immediately.

Acknowledgements. The author wishes to thank Drs. John Simmonds and Andrea Davidson for their critical reviews. ECORC contribution no. 02111.

References

Bailey MA, Boerma HR, Parrott WA (1993a) Genotype effects on proliferative embryogenesis and plant regeneration of soybean. In Vitro Cell Dev Biol 29P:102–108

Bailey MA, Boerma HR, Parrott WA (1993b) Genotype-specific optimization of plant regeneration from somatic embryos of soybean. Plant Sci 93:117–120

Barwale UB, Kerns HK, Widholm JM (1986) Plant regeneration from callus cultures of several soybean genotypes via embryogenesis and organogenesis. Planta 167:473–481

Bayliss MW (1980) Chromosomal variation in plant tissues in culture. In: Vasil IK (ed) Perspectives in plant cell and tissue culture. Int Rev Cytol, Suppl 11A. Academic Press, New York, pp 113–144

Buenrostro-Nava MT, Frantz HM, Ling PP, Finer JJ (1999) Animation and image analysis for evaluation of soybean (*Glycine max* L. Merrill.) somatic embryo growth. http://www.soygenetics.org/articles/sgn1999-015.htm

Cho M-J, Widholm JM, Vodkin LO (1995) Cassettes for seed-specific expression tested in transformed embryogenic cultures of soybean. Plant Mol Biol Rep 13:255–269

Chowrira GM, Akella V, Lurquin PF (1995) Electroporation-mediated gene transfer into intact nodal meristems in planta. Mol Biotechnol 3:17–23

Chowrira GM, Akella V, Fuerst PE, Lurquin PF (1996) Transgenic grain legumes obtained by in planta electroporation-mediated gene transfer. Mol Biotechnol 5:85–96

Christianson ML, Warnick DA, Carlson PS (1983) A morphologically competent soybean suspension culture. Science 222:632–634

Christou P (1990) Morphological description of transgenic soybean chimeras created by delivery, integration and expression of foreign DNA using electric discharge particle acceleration. Ann Bot 66:379–386

Christou P (1994) The biotechnology of crop legumes. Euphytica 74:165–185

Christou P, Swain WF, Yang N, McCabe DE (1989) Inheritance and expression of foreign genes in transgenic soybean plants. Proc Natl Acad Sci USA 86:7500–7504

Christou P, McCabe DE, Martinell BJ, Swain WF (1990) Soybean genetic engineering – commercial production of transgenic plants. TIBTECH 8:145–151

Cober E, Rioux S, Rajcan I, Donaldson PA, Simmonds DH (2002) Partial resistance to white mold in a transgenic soybean line. Crop Sci 43:92–95

Delannay X, Bauman TT, Beighley DH, Buettner MJ, Coble HD, DeFelice MS, Derting CW, Deidrick TJ, Griffin JL, Hagood ES, Hancock FG, Hart H, LaVallee BJ, Loux MM, Lueschen WE, Matson KW, Moots CK, Murdock E, Nickell AD, Owen MDK, Paschal II EH, Prochaska LM, Raymond PJ, Reynolds DB, Rhodes WK, Roeth FW, Sprankle PL, Tarochione LJ, Tinius CN, Walker RH, Wax LM, Weigelt HD, Padgette SR (1995) Yield evaluation of a glyphosate-tolerant line after treatment with glyphosate. Crop Sci 35:1461–1467

Dhir SK, Dhir S, Hepburn A, Widholm JM (1991) Factors affecting transient gene expression in electroporated *Glycine max* protoplasts. Plant Cell Rep 10:106–110

Di R, Purcell V, Collins GB, Ghabrial SA (1996) Production of transgenic soybean lines expressing the bean pod mottle virus coat protein precursor gene. Plant Cell Rep 15:746–750

Dinkins RD, Reddy MSS, Meurer CA, Yan B, Trick H, Thibaud-Nissen F, Finer JJ, Parrott WA, Collins GB (2001) Increased sulphur amino acids in soybean plants overexpressing the maize 15 kDa zein protein. In Vitro Cell Dev Biol Plant 37:742–747

Dinkins RD, Reddy MSS, Meurer CA, Redmond CT, Collins GB (2002) Recent advances in soybean transformation. In: Jaiwal PK, Singh RP (eds) Biotechnology of legumes. Springer, Berlin Heidelberg New York (in press)

Donaldson PA, Simmonds DH (2000) Susceptibility to *Agrobacterium tumefaciens* and cotyledonary node transformation in short-season soybean. Plant Cell Rep 19:478–484

Donaldson PA, Anderson T, Lane BG, Davidson AL, Simmonds DH (2001) Soybean plants expressing an active oligomeric oxalate oxidase from the wheat *gf-2.8* (germin) gene are resistant to the oxalate-secreting pathogen *Sclerotinia sclerotiorum*. Physiol Mol Plant Pathol 59:297-307

Finer JJ (1988) Apical proliferation of embryogenic tissue of soybean [*Glycine max* (L.) Merrill.]. Plant Cell Rep 7:238–241

Finer JJ, McMullen MD (1991) Transformation of soybean via particle bombardment of embryogenic suspension culture tissue. In Vitro Cell Dev Biol-Plant 27:175–182

Finer JJ, Nagasawa A (1988) Development of an embryogenic suspension culture of soybean (*Glycine max* Merrill). Plant Cell Tissue Organ Cult 15:125–136
Finer JJ, Vain P, Jones MW, McMullen MD (1992) Development of the particle inflow gun for DNA delivery to plant cells. Plant Cell Rep 11:323–328
Finer JJ, Cheng T-S, Verma DPS (1996) Soybean transformation: technologies and progress. In: Verma DPS, Shoemaker RC (eds) Soybean genetics, molecular biology and biotechnology. Biotechnology in agriculture series number 14. CAB International, Wallingford, UK, pp 249–262
Flavell RB (1994) Inactivation of gene expression in plants as a consequence of specific sequence duplication. Proc Natl Acad Sci USA 91:3490–3496
Gamborg OL, Miller RA, Ojima K (1968) Nutrient requirements of suspension cultures of soybean root cells. Exp Cell Res 50:151–158
Gritz L, Davies J (1983) Plasmid-encoded hygromycin B resistance: the sequence of hygromycin B phosphotransferase gene and its expression in *Escherichia coli* and *Saccharomyces cerevisiae*. Gene 25:179–188
Hadi MZ, McMullen MD, Finer JJ (1996) Transformation of 12 different plasmids into soybean via particle bombardment. Plant Cell Rep 15:500–505
Hammatt N, Davey MR (1987) Somatic embryogenesis and plant regeneration from cultured zygotic embryos of soybean (*Glycine max* L. Merr.). J Plant Physiol 128:219–226
Hansen G, Chilton M-D (1996) "Agrolistic" transformation of plant cells: integration of T-strands generated in plants. Proc Natl Acad Sci USA 93:14978–14983
Hazel CB, Klein TM, Anis M, Wilde HD, Parrott WA (1998) Growth characteristics and transformability of soybean embryogenic cultures. Plant Cell Rep 17:765–772
Hinchee MAW, Connor-Ward DV, Newell CA, McDonnell RE, Sato SJ, Gasser CS, Fischhoff DA, Re DB, Fraley RT, Horsch RB (1988) Production of transgenic soybean plants using *Agrobacterium*-mediated gene transfer. Bio/Technology 6:915–922
Horsch RB, Fry JE, Hoffman NL, Eichholtz D, Rogers SG, Fraley RT (1985) A simple and general method for transferring genes into plants. Science 227:1229–1231
Iida A, Yamashita T, Yamada Y, Morikawa M (1991) Efficiency of particle bombardment-mediated transformation is influenced by cell cycle stage in synchronized cultured cells of tobacco. Plant Physiol 97:1585–1587
Klein TM, Jones TJ (1999) Methods of genetic transformation: The gene gun. In: Vasil IK (ed) Molecular improvement of cereal crops. Advances in cellular and molecular biology of plants, vol 5. Kluwer, Dordrecht, pp 21–42
Klein TM, Wolf ED, Wu R, Sanford JC (1987) High-velocity microprojectiles for delivering nucleic acids into living cells. Nature 327:70–73
Komatsuda T, Ohyama K (1988) Genotypes of high competence for somatic embryogenesis and plant regeneration in soybean *Glycine max*. Theor Appl Gen 75:695–700
Kononov ME, Bassuner B, Gelvin SB (1997) Integration of T-DNA binary vector "backbone" sequences into the tobacco genome: evidence for multiple complex patterns of integration. Plant J 11:945–957
Lazzeri PA, Hildebrand DF, Collins GB (1985) A procedure for plant regeneration from immature cotyledon tissue of soybean. Plant Mol Biol Rep 3:160–167
Lazzeri PA, Hildebrand DF, Collins GB (1987) Soybean somatic embryogenesis: effects of nutritional, physical and chemical factors. Plant Cell Tissue Organ Cult 10:209–220
Lin W, Odell JT, Screiner RM (1987) Soybean protoplast culture and direct gene uptake and expression by cultured soybean protoplasts. Plant Physiol 85:856–861
Lippman B, Lippman G (1984) Induction of somatic embryos in cotyledonary node tissue of soybean, *Glycine max* L. Merr. Plant Cell Rep 3:215–218
Liu W, Torisky RS, McAllister KP, Avdiushko S, Hildebrand DF, Collins GB (1996) Somatic embryo cycling: evaluation of a novel transformation and assay system for seed-specific gene expression in soybean. Plant Cell Tissue Organ Cult 47:33–42
Maughan PJ, Philip R, Cho M-J, Widholm JM, Vodkin LO (1999) Biolistic transformation, expression and inheritance of bovine β-casein in soybean (*Glycine max*). In Vitro Cell Dev Biol Plant 35:344–349

McCabe D, Christou P (1993) Direct DNA transfer using electric discharge particle acceleration (ACCELL technology). Plant Cell Tissue Organ Cult 33:227–236

McCabe DE, Swain WF, Martinell BJ, Christou P (1988) Stable transformation of soybean (*Glycine max*) by particle acceleration. Bio/Technology 6:923–926

Meurer CA, Dinkins RD, Redmond CT, McAllister KP, Tucker DT, Walker DR, Parrott WA, Trick HN, Essig JS, Frantz HM, Finer JJ, Collins GB (2001) Embryogenic response of multiple soybean [*Glycine max* (L.) Merrill] cultivars across three locations. In Vitro Cell Dev Biol Plant 37:62–67

Moloney MM, Walker JM, Sharma KK (1989) High efficiency transformation of *Brassica napus* using *Agrobacterium* vectors. Plant Cell Rep 8:238–242

Murashige T, Skoog F (1962) A revised medium for rapid growth and bioassays with tobacco tissue cultures. Physiol Plant 15:473–498

Negrotto D, Jolley M, Beer S, Wenck AR, Hansen G (2000) The use of phosphomannose-isomerase as a selectable marker to recover transgenic maize plants (*Zea mays* L.) via *Agrobacterium* transformation. Plant Cell Rep 19:798–803

Oard JH, Paige DF, Simmonds JA, Gradziel TM (1990) Transient gene expression in maize, rice, and wheat cells using an airgun apparatus. Plant Physiol 92:334–339

Padgette SR, Kolacz KH, Delannay X, Re DB, LaVallee BJ, Tinius CN, Rhodes WK, Otero YI, Barry GF, Eichholtz DA, Peschke VM, Nida DL, Taylor NB, Kishore GM (1995) Development, identification, and characterization of a glyphosate-tolerant soybean line. Crop Sci 35: 1451–1461

Parrott WA, Dryden G, Vogt S, Hildebrand DF, Collins GB, Williams EG (1988) Optimization of somatic embryogenesis and embryo germination in soybean. In Vitro Cell Dev Biol 24: 817–820

Parrott WA, All JN, Adang MJ, Bailey MA, Boerma HR, Stewart Jr CN (1994) Recovery and evaluation of soybean plants transgenic for a *Bacillus thuringiensis* var. *kurstaki* insecticidal gene. In Vitro Cell Dev Biol 30P:144–149

Ponappa T, Brzozowski AE, Finer JJ (1999). Transient expression and stable transformation of soybean using jellyfish green fluorescent protein. Plant Cell Rep 19:6–12

Ranch JP, Oglesby L, Zielinski AC (1985) Plant regeneration from embryo-derived tissue cultures of soybeans. In Vitro Cell Dev Biol 21:653–657

Reddy MSS, Ghabrial SA, Redmond CT, Dinkins RD, Collins GB (2001) Resistance to *Bean pod mottle virus* in transgenic soybean lines expressing the capsid polyprotein. Phytopathology 91: 831–838

Register JC III, Peterson DJ, Bell PJ, Bullock WP, Evans IJ, Frame B, Greenland AJ, Higgs NS, Jepson I, Jiao S, Lewnau CJ, Sillick JM, Wilson HM (1994) Structure and function of selectable and non-selectable transgenes in maize after introduction by particle bombardment. Plant Mol Biol 25:951–961

Samoylov VM, Tucker DM, Parrott WA (1998a) Soybean [*Glycine max* (L.) Merrill] embryogenic cultures: the role of sucrose and total nitrogen content on proliferation. In Vitro Cell Dev Biol Plant 34:8–13

Samoylov VM, Tucker DM, Parrott WA (1998b) A liquid medium-based protocol for rapid regeneration from embryogenic soybean cultures. Plant Cell Rep 18:49–54

Sanford JC, Klein TM, Wolf ED, Allen N (1987) Delivery of substances into cells and tissues using a particle bombardment process. Part Sci Technol 5:27–37

Sanford JC, Devit MJ, Russell JA, Smith FD, Harpening PR, Roy MK, Johnston SA (1991) An improved helium driven biolistic device. Technique 3:3–16

Santarem ER, Finer JJ (1999) Transformation of soybean [*Glycine max* (L.) Merrill] using proliferative embryogenic tissue maintained on semi-solid medium. In Vitro Cell Dev Biol Plant 35: 451–455

Santarem ER, Pelissier B, Finer JJ (1997) Effect of explant orientation, pH, solidifying agent and wounding on initiation of soybean somatic embryos. In Vitro Cell Dev Biol Plant 33:13–19

Sato S, Newell C, Kalacz K, Tredo L, Finer J, Hinchee M (1993) Stable transformation via particle bombardment in two different soybean regeneration systems. Plant Cell Rep 12:408–413

Sautter C, Waldner H, Neuhaus-Url G, Galli A, Neuhaus G, Potrykus I (1991) Micro-targeting: high efficiency gene transfer using a novel approach for the acceleration of microprojectiles. Bio/Technology 9:1080–1085

Simmonds DH, Donaldson PA (2000) Genotype screening for proliferative embryogenesis and biolistic transformation of short-season soybean genotypes. Plant Cell Rep 19:485–490

Singh RJ, Klein TM, Mauvais CJ, Knowlton S, Hymowitz T, and Kostow CM (1998) Cytological characterization of transgenic soybean. Theor Appl Genet 96:319–324

Stewart CN Jr, Adang MJ, All JN, Boerma HR, Cardineau G, Tucker D, Parrott WA (1996) Genetic transformation, recovery and characterization of fertile soybean transgenic for a synthetic *Bacillus thuringensis cryIAc* gene. Plant Physiol 112:121–129

Sundaram P, Xiao W, Brandsma JL (1996) Particle mediated delivery of recombinant expression vectors to rabbit skin induces high titered polyclonal antisera (and circumvents purification of a protein immunogen). Nucleic Acid Res 24:1375–1377

Tian LN, Brown DCW, Voldeng H, Webb J (1994) In vitro response and pedigree analysis for somatic embryogenesis of long-day photoperiod adapted soybean. Plant Cell Tissue Organ Cult 36:269–273

Trick HN, Finer JJ (1998) Sonication-assisted *Agrobacterium*-mediated transformation of soybean [*Glycine max* (L.) Merrill] embryogenic suspension culture tissue. Plant Cell Rep 17: 482–488

Trick HN, Dinkins RD, Santarem ER, Di R, Samoylov V, Meurer CA, Walker DR, Parrott WA, Finer JJ, Collins GB (1997) Recent advances in soybean transformation. Plant Tissue Cult Biotechnol 3:9–26

Vain P, McMullen MD, Finer JJ (1993) Osmotic treatment enhances particle bombardment-mediated transient and stable transformation of maize. Plant Cell Rep 12:84–88

Walker DR, Parrott WA (2001) Effect of polyethylene glycol and sugar alcohols on soybean somatic embryo germination and conversion. Plant Cell Tissue Organ Cult 64:55–62

Walker DR, All JN, McPherson RM, Boerma HR, Parrott WA (2000) Field evaluation of soybean engineered with a synthetic *cry1Ac* transgene for resistance to corn earworm, soybean looper, velvetbean caterpillar (*Lepidoptera:* Noctuidae) and lesser cornstalk borer (*Lepidoptera:* Pyralidae). J Econ Entomol 93:613–622

Widholm JM (1995) In vitro selection and culture-induced variation in soybean. In: Verma DPS, Shoemaker RC (eds) Soybean genetics, molecular biology and biotechnology. Biotechnology in agriculture series number 14. CAB International, Wallingford, UK, pp 107–126

Wright MS, Launis KL, Novitzky R, Duesing JH, Harmes CT (1991) A simple method for the recovery of multiple fertile plants from individual somatic embryos of soybean [*Glycine max* (L.) Merrill]. In Vitro Cell Dev Biol 27P:153–157

Zhang Z, Xing A, Staswick P, Clemente TE (1999) The use of glufosinate as a selective agent in *Agrobacterium*-mediated transformation of soybean. Plant Cell Tissue Organ Cult 56:37–46

11 Genotoxic Effects of Tungsten Microparticles Under Conditions of Biolistic Transformation

J. BUCHOWICZ and C. KRYSIAK

11.1 Introduction

Integrative transformation depends on a temporal appearance of double-strand breaks (DSBs) in cellular DNA. Exogenously delivered foreign DNA should also have, or acquire soon after delivering, free ends to make recombination between the two DNA substrates possible. Hence, factors that induce DNA strand scission are of interest for biotechnological investigation. There is, however, no shortage of factors that are able to generate single-strand breaks (SSBs) or DSBs in native DNA. According to Citterio et al. (2000), even such ubiquitous agents as water, oxygen, and sunlight may be classified as genotoxins, i.e., substances or factors that can actively damage DNA. We refer to genotoxicity as being such DNA damage that is followed by detectable changes in DNA function (e.g., mutagenesis, carcinogenesis, impairments to morphogenetic program).

Interest in genotoxicity of tungsten has arisen from its uses for two different purposes, production of ammunition, and preparation of DNA delivering microprojectiles. Both military and biotechnological applications can result in internalization of metallic tungsten which, in addition to the immediate wounding, may be expected to show long-term metabolic effects. Some insight into the "top secret" military applications of tungsten and tungsten alloys may be found in a recent paper of Miller et al. (2001). Here, we are trying to summarize recent advances in knowledge relating to the biological significance of tungsten with a special reference to its interaction with native DNA and its genotoxicity to plants, as well as to the methods that allow the assessment of the damage to plasmid and cellular DNAs caused by metallic tungsten.

11.2 Biological Signi cance of Tungsten

11.2.1 Early Observations on Biological Effects of Tungsten

The first reports on possible biological functions of tungsten appeared in 1956 when Edwin S. Higgins completed his Ph.D. thesis on the biological role of molybdenum (Higgins 1956). It has become clear from his findings that tungstate is a dietary antagonist of molybdate in animal nutrition (Higgins et

Molecular Methods of Plant Analysis, Vol. 23
Genetic Transformation of Plants

al. 1956a) and competes with molybdate in *Aspergillus niger* (Higgins et al. 1956b). Similar competition between these two chemically related elements was then observed in bacteria (Bullen 1961) and higher plants (Notton and Hewitt 1971). Administration of sodium tungstate to rats was found to inhibit xanthine oxiase and other molybdenum-containing metalloenzymes, albeit with no impairment to the health of the animals (Johnson et al. 1974).

Although tungsten is still best known in its role as a molybdenum antagonist, it is also classified as an essential trace element for, at least, some organisms. Sodium tungstate was found to stimulate the growth of the thermophilic bacterium *Clostridium thermoaceticum* (Andreesen and Ljungdahl 1973) and to be essential for the hyperthermophilic archaeon *Pyrococcus furiosus* (Bryant and Adams 1989). Moreover, the radioactive tungsten isotope ^{185}W was shown to be incorporated into protein fractions of bacteria (Benemann et al. 1973), plants (Notton and Hewitt 1971), and animals (Wide et al. 1986). According to Adams (1993), some forms of life are tungsten-dependent.

11.2.2 Catalytic Activity of Simple Tungsten Compounds

The carbide WC was the first tungsten compound whose catalytic activity was detected. Levy and Boudart (1973) showed tungsten carbide to be an effective catalyst for many reactions that were earlier known to be readily catalyzed by platinum. Noticeably, no similar catalytic activity could be observed for elemental (metallic) tungsten. According to Bennett et al. (1974), the electronic density of states of tungsten carbide more nearly resembles that of platinum than that of tungsten.

More specific catalytic activities were observed for some organic compounds of tungsten. In particular, tungsten-alkylidene complexes, similarly to their molybdenum analogs, were found to catalyze alkene metathesis (Wengrovius et al. 1980), i.e., the cleavage and formation of carbon – carbon double bonds (Ivin and Mol 1997). In these catalytically active complexes, tungsten is covalently bound to a carbon atom of the organic component (Schaverien et al. 1986). According to Couturier et al. (1992), some neopentylidene-tungsten(VI) complexes are highly effective and stereoselective catalysts of the metathesis. Similar properties are expected from enzyme cofactors. It is, therefore, not surprising that an organic compound of tungsten (tungstopterin) has been identified as the prosthetic group of some metalloenzymes (Chan et al. 1995; Hu et al. 1999).

11.2.3 Tungstoenzymes

The first example of a tungsten-containing enzyme came from studies on the competition between tungsten and molybdenum for metal binding sites in proteins. Andreesen and Ljungdahl (1973) demonstrated that, in *C.*

thermoaceticum, tungstate stimulated the formation and activity of a formate dehydrogenase more efficiently than did molybdate. The next investigation step resulted in convincing evidence that tungsten is a component of catalytically active formate dehydrogenase (Ljungdahl and Andreesen 1975). Extensive purification of this dehydrogenase allowed Yamamoto et al. (1983) to demonstrate that it contains two tungsten atoms per four-subunit protein molecule.

Many molybdoenzymes are known to occur in higher plants (Kisker et al. 1997; Degtyarenko et al. 1999; Mendel and Schwarz 1999). There has been, however, no report on the replacement of molybdenum by tungsten in plant enzymes except for the already quoted paper of Notton and Hewitt (1971).

Typical tungstoenzymes should, probably, contain tungsten independently of its competition with molybdenum (neither in place nor in addition to molybdenum) even under conditions of excess molybdenum and just trace amounts of tungsten in soil or culture medium.

Metalloenzymes specific for tungsten are known from studies on a group of extremely thermophilic archaea which, according to Woese et al. (1990), belong to the most ancient organisms on the Earth. Well characterized preparations of typical tungstoenzymes were first isolated from *P. furiosus*. Using this organism as an experimental system, Mukund and Adams (1991) purified a red-colored tungsten-containing protein and identified it as an aldehyde ferrodoxin oxidoreductase. Further tungstoenzymes were isolated from other archaea (Schmitz et al. 1992; Johnson et al. 1993; Mukund and Adams 1993). Similarly to molybdopterin in molybdoenzymes, tungstopterin fulfils the role of the enzyme cofactor in tungstoenzymes (Johnson et al. 1993; Chan et al. 1995).

No tungstoenzyme has thus far been reported to occur in higher plants.

11.2.4 Tungsten—DNA Interaction

Heavy metal ions belong to the most extensively studied DNA damaging agents. They can induce both oxidative and hydrolytic DNA strand scission. Various structural forms of DNA are damaged. Detailed discussion on the mechanism of action and DNA damaging effects of heavy metals can be found in a number of excellent reviews (Barton 1986; Bianchi and Levis 1987; Chow and Barton 1992; Abrams and Murrer 1993; Cohen et al. 1993; OhHalloran 1993). Repair events, with a special reference to those occurring in higher plants, are described in detail by Britt (1996), Puchta and Hohn (1996), Gorbunova and Levy (1999), and Mengiste and Paszkowski (1999).

Neither tungstate nor tungsten ions have been included in the list of DNA damaging agents, so far. Metallic tungsten has been observed recently, however, to cause limited DNA damage.

When double-stranded, circular, covalently closed, supercoiled plasmid DNA is incubated with a suspension of tungsten microparticles, the intact DNA molecules disappear rapidly (Krysiak et al. 1999a). Simultaneously, nicked (open circular) DNA accumulates. Small amounts of linear DNA also arise, but no fragmentation products can be detected. The reaction rate depends on the amount of tungsten particles and, when the suspension density is high enough, the reaction is, apparently, completed within 20 min (Krysiak et al. 1999b). On the other hand, the reaction rate is not influenced by free radical scavengers or ascorbate. No changes in plasmid DNA structure can be observed when tungsten particles are replaced by tungsten particle leachates or commercial tungsten compounds (Mazus et al. 2000). Likewise, gold particles leave plasmid DNA intact. Plasmids relaxed by topoisomerases or linearized with restriction endonucleases remain unchanged during incubation with tungsten particles. Examples of electrophoretic separation of intact, relaxed, nicked, and linearized plasmids, including those treated with tungsten, are given in Section 11.4.1.

The specificity of tungsten-induced DNA damage may be summarized as an interaction between tungsten in the metallic state and native DNA in a compact, supercoiled conformation. Electrostatic forces are probably involved in this interaction and, as a result, tungsten-induced self-nicking of supercoiled DNA is observed.

11.3 Tungsten Microparticles in Biotechnological Applications

Similarly as in military applications, tungsten is used for biotechnological purposes because of its very high specific gravity and extreme hardness. These properties allow tungsten microparticles to penetrate target tissues for a desired depth with minimal mechanical injury. Therefore, conditions were found to coat them with DNA constructs or other biologically active materials and to deliver the resulting microprojectiles into living cells (see Sect. 11.3.1.2).

Another advantage of tungsten particles results from their easy availability and relatively low prices. They arise from the reduction of tungsten(VI) oxide to elemental tungsten with hydrogen at 1000°C, i.e., much below the melting point (3410°C) of this metal. The resulting metallic powder consists of particles that are irregularly shaped and very heterogeneous in size. A more detailed characterization of tungsten metal particles may be found in a review by Sanford (1988).

11.3.1 Biolistic Transformation

11.3.1.1 An Overview

The best-known biotechnological application of tungsten is related to its use as the DNA carrier in biolistic transformation of crop plants. This method of genetic transformation, known also as the biolistic process (Sanford 1988) or particle bombardment (Klein et al. 1992), was used for the first time in the late 1980s to deliver RNA and DNA into epidermal tissue of onion (Klein et al. 1987). The newly invented technique has since been widely used in studies on transient and stable transformation of various plant species. The results of early investigations on this topic are summarized by Christou et al. (1990), Klein et al. (1992), and Sanford et al. (1993).

As the use of tungsten particles in transformation experiments has been continued, some disadvantages of this DNA carrier have become apparent. According to Christou et al. (1990), tungsten is readily coated with a thin film of oxide which may be detrimental to plant cells. Acidification of culture media and other (unspecified) reasons for tungsten cytotoxicity were reported by Russell et al. (1992). It was also announced that tungsten can catalytically degrade DNA (Sanford et al. 1993). The use of gold, instead of tungsten, was therefore recommended. However, many workers have not followed this advice.

While the early reports on tungsten particle-mediated transformation are included in the above-quoted reviews, further data on the same topic (with the omission of transient gene expression) are summarized in Table 11.1. Notably, all the main cereal species and many other crop plants have been stably transformed with DNA-coated tungsten particles. DNA constructs to be integrated and to function in either nuclear or plastid genome have been prepared and successfully used. Bacterial sequences coding for β-glucuronidase (GUS) and phosphinotricin acetyltransferase have been most frequently used as reporter and marker genes, respectively. Further transgenes have been used to improve crop quality and productivity. In particular, *Bacillus thuringiensis* genes coding for insecticide toxins have been successfully integrated into the genomes of maize and tobacco. Genes increasing the resistance to viral infections have been introduced via particle bombardment into maize and chrysanthemum. Transformation of rice with a barley gene *HVA1* resulted in its increased tolerance to water deficit and salt stress (for references, see Table 11.1).

Although the lists of similar examples, common to tungsten and gold DNA-carriers indiscriminately, are much longer (see reviews by Vasil 1994; Christou 1997; Tyagi and Mohanty 2000; Ingram et al. 2001; Kleter et al. 2001; Kuiper et al. 2001; Patnaik and Khurona 2001), there has been no report indicating that a plant species can be transformed with DNA-coated gold, but not tungsten, particles.

In addition to higher plants, many other organisms have been reported to be stably transformed using DNA-coated tungsten particles. These are *Escherichia coli* (Smith et al. 1992), *Saccharomyces cerevisiae* (Butow et al. 1996), *A. nidulans* (Herzog et al. 1996), *Chlamydomonas reinhardtii* (Boyton and Gillham 1993), *Dictyostelium discoideum* (Wetterauer et al. 2000), the diatom *Phaeodactylum tricornutum* (Apt et al. 1996) and, as reviewed by Klein et al. (1992), vertebrate cells, tissues and organs.

Table 11.1. Stable transformation of higher plants with DNA-coated tungsten particles. (Summarized are reports that have appeared since 1993; for earlier data, see Sanford et al. 1993)

Plant species	Target tissue	Transgene[a]	Reference
Monocots			
Avena sativa	Callus	*gus, bar*	Pawlowski and Sommers (1998)
Dendrobium sp.	Callus	*luc*	Chia et al. (1994)
Hordeum vulgare	Cell suspension	*gus, hpt*	Hensgens et al. (1993)
Lolium perenne	Cell suspension	*gus, hpt*	van der Maas et al. (1994)
Oryza sativa	Cell suspension	*gus, hpt*	Hensgens et al. (1993)
	Cell suspension	*HVA1*	Xu et al. (1996)
Pennisetum glaucum	Callus	*gus, hpt*	Lambe et al. (1995)
Triticum aestivum	Leaf[b]	*gus, cat*	Daniell (1993)
	Callus	*gus, bar*	Dobrzanska et al. (1997)
Zea mays	Cell suspension	*nptII, MDMVcp*	Murry et al. (1993)
	Cell suspension	*gus, bar*	Buising and Benbow (1994)
	Cell suspension	*gus, nptII*	Register et al. (1994)
	Callus	*bar, Bt cry*	Moellenbeck et al. (2001)
	Callus	*bar*, modified *p1*	Cocciolone et al. (2001)
Dicots			
Asparagus of cinalis	Callus	*gus, bar*	Cabrera-Ponce et al. (1997)
Beta vulgaris	Cell suspension[b]	*gus, cat*	Daniell (1993)
Brassica napus	Callus	*gus, nptII*	Chen and Beversdorf (1994)
Catharanthus roseus	Cell suspension	*gus*	van der Fits and Memelink (1997)
Chrysanthemum sp	Callus	*nptII, TSWVcp*	Yepes et al. (1995)
Galega orientalis	Callus	*gus, nptII*	Collen and Jarl (1999)
Glycine max	Cell suspension	*gus, hpt*	Cho et al. (1995)
Nicotiana tabacum	Cell suspension[b]	*gus, cat*	Daniell (1993)
	Leaf	*gus, bar*	Buising and Benbow (1994)
	Leaf[b]	*gus, aadA*	Zoubenko et al. (1994)
	Leaf[b]	*aadA, EPSPS*	Daniell et al. (1998)
	Leaf[b]	*aadA, Bt cry*	De Cosa et al. (2001)
Phaseolus vulgaris	Seedling apex	*gus*	Cruz de Carvalho et al. (2000)
Vitis sp.	Cell suspension	*gus, nptII*	Herbert et al. (1993)

[a] *aad*A, aminoglycoside 3′-adenylyltransferase; *bar*, phosphinotricin acetyltransferase; *Bt cry*, *Bacillus thuringensis* Cry protein; *cat*, chloramphenicol acetyltransferase; *EPSPS*, 5-enol-pyruvyl shikimate-3-phosphate synthase; *gus* (known also as *uid*A), β-glucuronidase; *hpt*, hygromycin phosphotransferase; *HVA1*, *Hordeum vulgare* LEA protein; *luc*, luciferase; *MDMV*cp, maize dwarf mosaic virus coat protein; *npt*II, neomycin phosphotransferase II; *p1*, *Myb*-homologous regulatory protein; *TSWV*cp, tomato spotted wilt virus coat protein.
[b] Transgene integrated into the plastid genome.

11.3.1.2 Technical Details

The initial and final steps of plant transformation, i.e., the preparation of a desired DNA construct (usually in the form of a double-stranded, circular plasmid) and regeneration of transgenic plants from the transformed cell or explant, are common to all methods of stable transformation. Two other steps, on the other hand, are typical of the biolistic method. These are the coating of metal particles with DNA and bombardment of a target with such microprojectiles.

When DNA is precipitated with polyamines in the presence of metal particles (no difference between tungsten or gold), the DNA precipitate is physically bound to these particles. The nature of this adhesion or adsorption, often referred to as coprecipitation, has not been studied in detail. A complete coating protocol is given in Sanford et al. (1993). Here, we describe a somewhat simplified version of this protocol. We have used it in our laboratory to transform wheat (Dobrzanska et al. 1997) with the plasmid pAHC25, constructed and described by Christensen and Quial (1996).

Commercial tungsten particles (0.2–1.5 μm in diameter) are washed with distilled water, sterilized with 70% (v/v) ethanol and suspended (50 mg/ml) in 50% (v/v) glycerol. To a 50-μl portion of the glycerol suspension (2.5 mg of tungsten), 5 μl of plasmid DNA solution (1 μg/μl in 10 mM Tris-HCl, 1 mM EDTA, pH 8.0), 50 μl of 2.5 M $Ca(NO_3)_2$, and 20 μl of 0.1 M spermidine are added. After 3 min of vortexing at room temperature, the mixture is left for 5 min in an ice-bath and centrifuged (1000× g, 5 min, room temperature). The pellet (DNA-coated particles) is washed with 70% ethanol, suspended in 99% ethanol, divided into 10-μl portions (ca. 0.5 mg of tungsten and nearly 1 μg of DNA), and used for bombardment without delay.

When immature wheat embryos are subjected to bombardment, a gas pressure of 1350 psi (9.3 MPa) and a 6-cm target distance (helium-driven gene gun PDS-1000/He) seem to be optimal. Usually, one shot (10 μl of the ethanolic suspension, see above) is given per group of 20 embryos. Embryos are sterilized (70% ethanol, 7% calcium hypochlorite) prior to bombardment and left under sterile conditions for plant regeneration.

11.3.2 Biolistic Inoculation and Related Applications of Tungsten Particles

Particle bombardment was found to be a useful alternative method of inoculation in studies on plant infection with RNA and DNA viruses.

The study of tobacco mosaic virus (TMV) has been aided by the production of cDNA clones for molecular manipulation and analysis (Boyer and Haenni 1994). However, low infectivity following the manual inoculation of plants using cDNA-based constructs has been observed (Ding et al. 1995). For this reason, Dagless et al. (1997) developed an alternative method. Tungsten particles coated with a viral cDNA-derived construct are bombarded into tobacco leaves using a commercially available gene gun. Contrary to the lack of infection via manual inoculation, the biolistic procedure results in the infection of all TMV host plants tested. According to the interpretation given, the new method is highly effective as it allows the plasmid vector to be delivered directly into the nuclei.

The same technique was used in studies on the host range of geminiviruses. To increase infectivity of a viral DNA preparation, Paximadis et al. (1999) condensed it onto tungsten particles under typical coating conditions. The DNA-coated particles were delivered then into plant leaves with a helium-driven gene gun. Symptoms of the biolistically inoculated test plants (tobacco and other) were identical to those observed for field-grown and naturally infected plants.

Another example of the usefulness of tungsten particles in biological investigations may be illustrated by a recent paper of Schweizer et al. (2000). This time, tungsten particles were coated with double-stranded RNA and bombarded into leaf segments of wheat and other cereals. A sequence-specific interference of double-stranded RNA with gene function was observed. Similarly, biolistic introduction of promoterless DNA constructs was found to cause post-transcriptional gene silencing in transgenic tobacco (Palauqui and Balzergue 1999).

11.4 Assessment of Tungsten-Induced DNA Lesions

11.4.1 Electrophoretic Analysis of Tungsten-Damaged Plasmid DNA

As outlined above (Sect. 11.3.1), DNA-coated tungsten particles are routinely used for genetic transformation of crop plants. Appropriate DNA constructs are usually propagated in bacterial cells in the form of plasmid DNA. The most commonly used method of plasmid DNA isolation results in preparations consisting of double-stranded, circular, covalently closed, negatively supercoiled DNA molecules (Sambrook et al. 1989). Such DNA preparations are referred to as intact or native plasmid DNA.

Enzymatic (and, conceivably, tungsten-catalyzed) degradation of intact plasmid DNA may lead to its relaxation, nicking, linearization, and fragmentation. The degradation products can be separated by conventional gel electrophoresis. However, when the electrophoretic separation is carried out in the presence of ethidium bromide (EtBr), relaxed and supercoiled plasmid forms can hardly be separated from each other. On the other hand, when the separation is done in the absence of EtBr and this intercalator is used only to visualize the DNA products, relaxed (but still circular covalently closed) and nicked (open circular) forms of the same plasmid appear as a single band. Hence, when there is a need to separate the relaxed form from both nicked and supercoiled forms, two samples of the same plasmid preparation should be analyzed in parallel, in the presence and in the absence of EtBr. The corresponding electrophoretic patterns are schematically shown in Fig. 11.1. Some further details are given in Figs. 11.2–11.4, where the separation of DNA products arising from the intact plasmid pAHC25 under various conditions is illustrated.

A typical example of tungsten-induced changes in electrophoretic mobility of pAHC25 is shown in Fig. 11.2. The rapidly moving supercoiled form (main

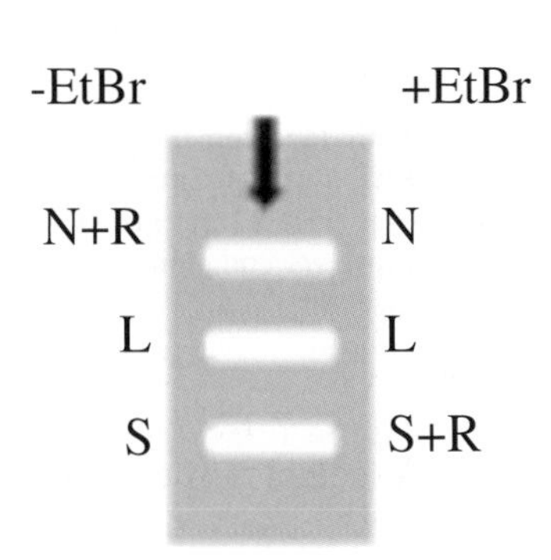

Fig. 11.1. Electrophoretic separation of various forms of plasmid DNA in agarose gels in the presence and absence of ethidium bromide (*EtBr*). The scheme is based on the original paper of Johnson and Grossman (1977) and our own experience (see Figs. 11.2–11.4). *S* Supercoiled, *R* relaxed, *N* nicked (open circular), *L* linear

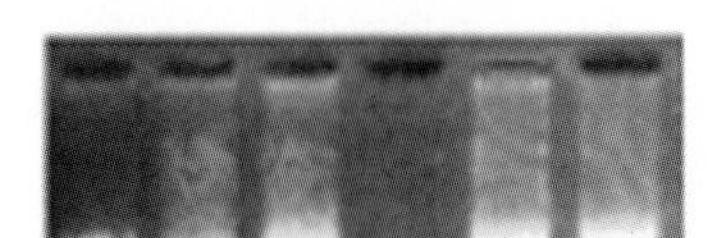

Fig. 11.2. Effects of tungsten particles and topoisomerase I on the integrity of the plasmid pAHC25. *Lane 1* Untreated plasmid; *lane 2* plasmid incubated with tungsten particles (5 µg DNA, 250 µg tungsten, 50 µl TE buffer, pH 7.4, 21°C, 20 min); *lane 3* plasmid linearized with *Sma*I; *lane 4* plasmid relaxed with topoisomerase; *lane 5* plasmid relaxed with topoisomerase and incubated with tungsten particles; *M* size markers (λ DNA digested with *Hin*dIII, linear fragments of 23.1, 9.4, 6.6, 4.3, 2.3 and 2.0 kb in size). Electrophoresis was in 1% agarose gel, ethidium bromide was used only to visualize DNA products. (Krysiak 2002)

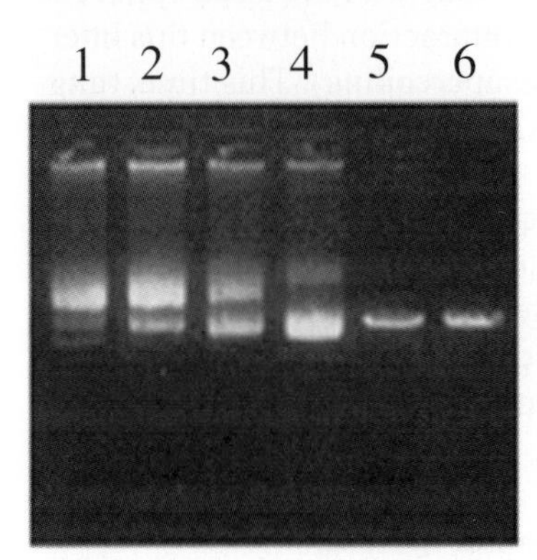

Fig. 11.3. Time course of tungsten-promoted DNA damage. Intact (*lanes1—4*) and linearized (*lanes 5* and *6*) pAHC25 preparations were incubated with tungsten particles (see Fig. 11.2 legend) for various time periods. *Lanes 1* and *6* 60 min, *lane 2* 30 min, *lane 3* 15 min, *lanes 4* and *5* 0 min. (Krysiak 2002)

component of freshly prepared plasmid DNA) disappears almost completely and a slowly moving product (relaxed or apparently relaxed plasmid form) predominates on the electrophoregram. In addition, small amounts of a product corresponding to linearized pAHC25 appear. Similar changes are observed when tungsten particles are replaced by topoisomerase I. The tungsten-promoted changes are time-dependent (Fig. 11.3, lanes 1–4) and

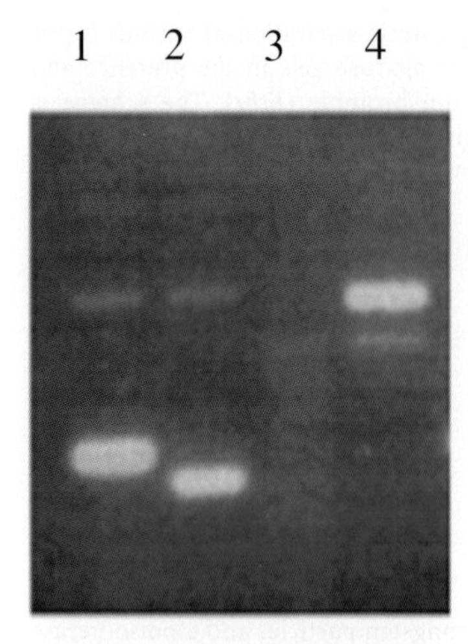

Fig. 11.4. Electrophoretic identification of nicked (open circular) DNA as the main product of the damage of pAHC25 by tungsten particles. *Lane 1* Untreated plasmid; *lane 2* plasmid digested with topoisomerase I; *lane 3* plasmid linearized with *Sma*I; *lane 4* plasmid incubated with tungsten particles (see Fig. 11.2 legend). Electrophoresis was in 1% agarose gel containing ethidium bromide (0.6 µg/ml). (Krysiak 2002)

show some substrate specificity. While the intact plasmid disappears rapidly on incubation with tungsten particles, plasmid preparations relaxed with topoisomerase (Fig. 11.2, lane 5) or linearized with *Sma*I (Fig. 11.3, lane 6) are, apparently, not damaged. Therefore, it could seem justified to conclude tentatively that the effects of tungsten and topoisomerase are, essentially, the same. The incorrectness of this conclusion became apparent after the electrophoretic analysis was carried out in the presence of EtBr (interaction between this intercalator and relaxed plasmid NA results in its resupercoiling). This time, tungsten-damaged and topoisomerase-treated pAHC25 preparations moved at completely different rates (Fig. 11.4).

It is, therefore, clear that tungsten particles cause DNA nicking, not its true relaxation. Rapid nicking of the circular double helix may, conceivably, result in its linearization. However, no further tungsten-promoted DNA strand scission seems to be possible after such structural changes as relaxation (nicking-closing), apparent relaxation (nicking) or linearization (double nicking).

11.4.2 A Modi ed TUNEL Method for Detection of Cellular DNA Fragmentation

It is difficult to apply electrophoresis for the analysis of DNA integrity in tissues bombarded with tungsten particles. Post-bombardment DNA nicking in living cells and, especially, artifactual DNA degradation in the presence of tungsten particles (even under conditions of pulsed field gel electrophoresis) make it impossible to prepare cellular DNA for analytical purposes. Clearly, there is a need to use an indirect method. Among these, the TUNEL (terminal deoxynu-

Table 11.2. Effect of bombardment of wheat embryos with tungsten particles on template-independent end labeling of nuclear DNA. (Adapted from Krysiak et al. 1999b)

Embryos[a]	DNA content (μg)	DNA radioactivity (cpm/μg)
Bombarded	31.7 ± 3.2	33.4 ± 2.9
Nonbombarded	34.0 ± 1.5	20.4 ± 1.7

[a] Mature wheat embryos were isolated as described elsewhere (Krysiak et al. 1999b), germinated at 21°C for 6h and subjected to bombardment with uncoated (no DNA) tungsten particles (for conditions, see Sect. 11.3.1). A crude nuclear fraction (900× g pellet) was prepared and tested under conditions of a modified TUNEL method. The complete mixture (0.1 ml) contained 200 mM sucrose, 100 mM sorbitol, 37 mM Tris-HCl (pH 8.0), 30 mM KCl, 12 mM $MgCl_2$, 7 mM EDTA, 5 mM dithiothreitol, 1.5 mM $CaCl_2$, 1 mM ATP, 1 μCi of [α-^{32}P]dCTP (400 Ci/mmol), and approximately 10^6 freshly prepared nuclei (34 μg DNA). After incubation for 20 min at 21°C, the nuclei were spun down (900 ×g, 6 min, 4°C), suspended in 1 ml of incubation mixture in which [^{32}P]dCTP had been replaced by dCTP and pelleted again. Total DNA was then isolated from the pellet with a phenol method. Nonbombarded embryos were similarly germinated and used to prepare a nuclear fraction for incubation with [^{32}P]dCTP. The amounts and radioactivities of DNA in nuclear pellets of bombarded and nonbombarded embryos are compared. Triplicate experiments, mean ± SD

cleotidyl transferase-mediated dUTP nick end-labeling) method seems to look promising enough. Successful use of this method in studies on nuclear DNA fragmentation during programmed cell death (apoptosis) in plant tissues has been reported repeatedly (Ryerson and Heath 1996; Wang et al. 1996; Balk and Leaver 2001; Yao et al. 2001). However, the method is laborious and requires expensive materials and equipment.

To avoid these inconveniences, we have developed a simplified version of the TUNEL method. We have taken advantage of the fact that wheat embryos show a high terminal deoxynucleotidyl transferase (TdT)-like activity (Brodniewicz-Proba and Buchowicz 1980).The plant enzyme is undistinguishable from the Bollum-type mammalian TdT with respect to its catalytic activity (although it does not prefer cacodylate to Tris buffer). Hence, to detect template-independent end labeling of DNA in wheat, the use of a commercial TdT preparation is unnecessary. Further, instead of fluorescently labeled dUTP, we used [^{32}P]dCTP to label free 3′-OH ends of nuclear DNA. A complete protocol is given in Table 11.2, where quantitative data on the DNA end labeling in bombarded and control wheat embryos are also presented. It is evident that DNA of bombarded embryos is labeled more extensively that that of nonbombarded ones. Therefore, it may be assumed that DSBs appear in cellular DNA as a result of particle bombardment. A similar conclusion was reached recently by Abranches et al. (2000) who applied a combination of fluorescence in situ hybridization (FISH) and three-dimensional confocal microscopy in studies on transgene integration sites in wheat callus bombarded with DNA-coated gold particles.

11.5 Post-Bombardment Inhibition of Somatic Embryogenesis

The effect of particle bombardment on cell totipotency is of primary importance for somatic embryogenesis, morphogenesis and plant regeneration. Data on this topic are, however, rather scant, although the biolistic technique is widely used in plant transformation which, typically, requires regeneration of a whole plant from a single transformed cell.

Attempts to optimize biolistic conditions showed that, while it is relatively easy to achieve nearly 100% efficiency of transient transgene expression, the efficiency of stable transformation is always much lower. Depending on the experimental conditions, and the manner of calculation in particular, the yield of stable transformation has varied from 0.1 to approximately 10%. The highest losses in stable transformation have been usually observed at an early stage of plant transformation. For example, bombardment of immature wheat embryos with DNA-coated gold particles reduces their ability to give regenerants by up to 50% (Vasil et al. 1993). When tungsten particles are used for the same purpose, the embryogenic potential of wheat embryos is reduced to a similar extent (Dobrzanska et al. 1997). The decrease of the regeneration capability can be caused by mechanical damage to cellular DNA by the intruding metal particles and, in the case of tungsten particles, by the post-bombardment presence of this metal in transformed cells. In addition, unintentional side effects of foreign genes introduced together with metal particles are possible.

Transgene expression can, sometimes, interfere with molecular events that are essential for somatic embryogenesis and morphogenesis. For example, expression of the jellyfish gene coding for green fluorescent protein (GFP), often used as a reporter gene in transgenic plants (Haseloff and Amos 1995), can generate free radicals which, in turn, damage genomic DNA. Plasmid vector sequences can, in addition, cause gene rearrangements (Fu et al. 2000; Labra et al. 2001). As reviewed by Kuiper et al. (2001), many foreign genes delivered to crop plants to improve their quality appear to have unexpected secondary effects. Moreover, expression of transgenes can cause post-transcriptional silencing of structurally related endogenous genes (Meyer and Saedler 1996; Meins 2000; Mittelstein Scheid and Paszkowski 2000; Muskens et al. 2000). Unrelated transgenes can, on the other hand, induce apoptosis (Mittler and Rizhsky 2000).

To observe the effect of particle bombardment itself, not obscured by possible effects of simultaneously delivered DNA, experiments with the use of uncoated particles should be carried out. The data in Table 11.3 allow us to compare the effects of DNA-coated and uncoated tungsten particles on somatic embryogenesis in callus initiated from immature wheat embryos. It is evident that both coated and naked tungsten particles cause a significant decrease in the ability of embryos to give embryogenic callus. Although the efficiency of callus initiation is not influenced by bombardment, the frequency of embryogenic events in calli derived from bombarded (with either coated or uncoated particles) embryos is lower by half than that in callus derived from

Table 11.3. Effect of particle bombardment on somatic embryogenesis. (Adapted from Krysiak et al. 1999b)

Specification of experiment stage[a]	Biolistic treatment		
	W+DNA	W	None
Number of ZEs used	400	420	360
Number of calli initiated	376	380	311
Number of SEs formed[b]	2271	2357	3929
Average number of SEs per ZE[b]	5.7	5.6	10.9
Relative frequency of SE formation[b]	52.3	51.4	100
Number of SEs used	45	45	45
Number of germinating SEs	34	33	42
Number of transgenic regenerants[c]	4	0	0

[a] Immature zygotic embryos of wheat (*Triticum aestivum* L.), cv. Chinese Spring, were bombarded with either DNA-coated (W + DNA) or naked (W) tungsten particles. Coating and bombardment conditions are given in Section 11.3.1. Plasmid pAHC25 (carrying *gus* and *bar* genes) was used as the DNA construct. Bombarded and nonbombarded embryos were subjected to in vitro culture conditions leading to initiation of callus and induction of somatic embryogenesis. ZE, Zygotic embryo; SE, somatic embryo.
[b] Differences between figures of column 2 vs. 4 and column 3 vs. 4 are statistically significant (Student t-test, $p < 0.05$).
[c] These can grow in the presence of phosphinotricin and show GUS activity (growth and test conditions were the same as in Dobrzanska et al. 1997).

nonbombarded embryos. In addition to the inhibition of somatic embryogenesis, subsequent plant regeneration was also observed to be less satisfactory in bombarded than in control tissues (Krysiak et al. 1999b).

The reason for which tungsten particles inhibit somatic embryogenesis may be related to their ability to cause DNA nicking. Quite conceivably, tungsten-promoted strand scission can interfere with chromosomal DNA condensation. The chromosome condensation is a prerequisite for correct DNA partitioning during cell division (Holmes and Cozzarelli 2000) and rapidly proliferating cells are characteristic of the earliest stages of somatic embryo formation in most explants (Zimmerman 1993; Maheshwari et al. 1995; Kemper et al. 1996; Yasuda et al. 2001). Artificially induced DSBs are also known to trigger apoptosis, even in tissues consisting of rapidly proliferating cells (Rich et al. 2000).

On the other hand, occasional generation of DSBs in chromosomal DNA may be essential for transgene integration. According to Mengiste and Paszkowski (1999) and Hanin et al. (2000), DSBs, referred to by them as genotoxic stress, are highly recombinogenic and, hence, can facilitate integrative transformation. Likewise, linearization of circular plasmids by tungsten may promote recombination between the delivered DNA construct and cellular DNA. It is, therefore, not excluded that in some biotechnological applications tungsten may be superior to gold.

11.6 Concluding Remarks

Interest in tungsten genotoxicity is related to its uses in biolistic techniques. Direct physical contact between native DNA and tungsten microparticles under coating and post-bombardment conditions can cause limited DNA strand scissions (SSBs and DSBs in supercoiled DNA). The metallic state of tungsten and the torsionally strained DNA structure seem to be essential for the tungsten–DNA interaction. The tungsten-induced DNA damage is followed by a selective inhibition of somatic embryogenesis.

Simple methods for the detection of subtle structural changes in plasmid DNA are available. Conventional gel electrophoresis in the presence and, in parallel, in the absence of ethidium bromide allows the detection of a single nick within double-stranded circular DNA. Assessment of the tungsten-induced cellular DNA damage is more difficult. Nevertheless, fragmentation of nuclear DNA can easily be detected with a modified TUNEL method.

Contrary to tungsten, gold causes no DNA nicking. Hence, to avoid undesirable side effects of tungsten, DNA-coated gold particles are most commonly used for plant transformation. Thus, as far as biotechnology is concerned, the problem is solved. For molecular biology, on the other hand, challenging questions have just been asked. Why tungsten particles inhibit cell differentiation (somatic embryogenesis), but remain without effect on proliferative cell division? What is the role of the metallic state in the tungsten–DNA interaction? And why, upon contact with the surface of tungsten microparticles, plasmids behave like a self-cleaving DNA? The nature of interaction between native DNA and tungsten metal particles may be expected to become a topic of increasing interest.

References

Abrams MJ, Murrer BA (1993) Metal compounds in therapy and diagnosis. Science 261:725–730

Abranches R, Santos AP, Wegel E, Williams S, Castilho A, Christou P, Shaw P, Stoger E (2000) Widely separated multiple transgene integration sites in wheat chromosomes are brought together at interphase. Plant J 24:713–726

Adams MWW (1993) Enzymes and proteins from organisms that grow near and above 100°C. Annu Rev Microbiol 47:627–658

Andreesen JR, Ljungdahl LG (1973) Formate dehydrogenase of *Clostridium thermoaceticum*: incorporation of selenium-75, and the effects of selenite, molybdate, and tungstate on the enzyme. J Bacteriol 116:867–873

Apt KE, Kroth-Pancic PG, Grossman AR (1996) Stable nuclear transformation of the diatom *Phaeodactylum tricornutum*. Mol Gen Genet 252:572–579

Balk J, Leaver CJ (2001) The PET1-CMS mitochondrial mutation in sunflower is associated with premature programmed cell death and cytochrome release. Plant Cell 13:1803–1818

Barton JK (1986) Metals and DNA: molecular left-handed complements. Science 233:727–734

Benemann JR, Smith GM, Kostel PJ, McKenna CE (1973) Tungsten incorporation into *Azotobacter vinelandi* nitrogenase. FEBS Lett 29:219–221

Bennett LH, Cuthill JR, McAlister AJ, Erickson NE (1974) Electronic structure and catalytic behavior of tungsten carbide. Science 184:563–565
Bianchi V, Levis A (1987) Recent advances in chromium genotoxicity. Toxicol Environ Chem 15:1–24
Boyer J-C, Haenni A-L (1994) Infectious transcripts of cDNA clones of RNA viruses. Virology 198:415–426
Boyton JE, Gillham NW (1993) Chloroplast transformation in *Chlamydomonas*. Methods Enzymol 217:510–536
Britt AB (1996) DNA damage and repair in plants. Annu Rev Plant Physiol Plant Mol Biol 47:75–100
Brodniewicz-Proba T, Buchowicz J (1980) Properties of a terminal deoxynucleotidyl transferase isolated from wheat germ. Biochem J 191:139–145
Bryant FO, Adams MWW (1989) Characterization of dehydrogenase from the hyperthermophilic archaebacterium, *Pyrococcus furiosus*. J Biol Chem 264:5070–5079
Buising CM, Benbow RM (1994) Molecular analysis of transgenic plants generated by microprojectile bombardment: effect of petunia transformation booster sequence. Mol Gen Genet 243:71–81
Bullen WA (1961) Effect of tungstate on the uptake and function of molybdate in *Azotobacter agilis*. J Bacteriol 82:130–134
Butow RA, Henke RM, Moran JV, Belcher SM, Perlman PS (1996) Transformation of *Saccharomyces cerevisiae* mitochondria using the biolistic gun. Methods Enzymol 264:265–278
Cabrera-Ponce JL, Lopez L, Assad-Garcia N, Medina-Arevalo C, Bailey AM, Herrera-Estrella L (1997) An efficient particle bombardment system for the genetic transformation of asparagus (*Asparagus of cinalis* L.). Plant Cell Rep 16:255–260
Chan MK, Mukund S, Kletzin A, Adams MWW, Rees DC (1995) Structure of a hyperthermophilic tungstopterin enzyme, aldehyde ferrodoxin oxidoreductase. Science 267:1463–1469
Chen JL, Beversdorf WD (1994) A combined use of microprojectile bombardment and DNA imbibition enhances transformation efficiency of canola (*Brassica napus* L). Theor Appl Genet 88:187–192
Chia T-F, Chan Y-S, Chua N-H (1994) The firefly luciferase gene as a non-invasive reporter for *Dendrobium* transformation. Plant J 6:441–446
Cho M-J, Widholm JM, Vodhin LD (1995) Cassettes for seed-specific expression tested in transformed cultures of soybean. Plant Mol Biol Rep 13:255–269
Chow CS, Barton JK (1992) Transition metal compounds as probes of nucleic acids. Methods Enzymol 212:219–242
Christensen AH, Quail PH (1996) Ubiquitin promoter-based vectors for high-level expression of selectable and/or screenable marker genes in monocotyledonous plants. Transgenic Res 5:213–218
Christou P (1997) Rice transformation: bombardment. Plant Mol Biol 35:197–203
Christou P, McCabe DE, Martinell BJ, Swain WF (1990) Soybean gene engineering: commercial production of transgenic plants. Trends Biotechnol 8:145–151
Citterio E, Vermeulen W, Hoeijmakes HJ (2000) Transcriptional healing. Cell 101:447–450
Cocciolone SM, Chopra S, Flint-Garcia SA, McMullen MD, Peterson T (2001) Tissue-specific patterns of a maize *Myb* transcription factor are epigenetically regulated. Plant J 27:467–478
Cohen MD, Korgacin B, Klein CB, Costa M (1993) Mechanisms of chromium carcinogeneticity and toxicity. Crit Rev Toxicol 23:255–281
Collen AMC, Jarl CI (1999) Comparison of different methods for plant regeneration and transformation of the legume *Galega orientalis* Lam. (goats rue). Plant Cell Rep 19:13–19
Couturier JL, Paillet C, Leconte M, Basset J-M, Weiss K (1992) A cyclometalated aryloxy(chloro)neopentylidene-tungsten complexes: a highly active and stereoselective catalyst for the metathesis of *cis-* and *trans*-2-pentene, norbornene, 1-methylnorbornene, and ethyl oleate. Angew Chem Int Ed Engl 31:628–631
Cruz de Carvalho MH, Le BV, Zuilly-Fodil Y, Pham Thi AT, Tran Thahn V (2000) Efficient whole plant regeneration of common bean (*Phaseolus vulgaris* L.) using thin-cell-layer culture and silver nitrate. Plant Sci 159:223–232

Dagless EM, Shintaku MH, Nelson RS, Foster GD (1997) A CaMV promoter driven cDNA clone of tobacco mosaic virus can infect host plant tissue despite being uninfectious when manually inoculated onto leaves. Arch Virol 142:183–191

Daniell H (1993) Foreign gene expression in chloroplasts of higher plants mediated by tungsten particle bombardment. Methods Enzymol 217:536–550

Daniell H, Datta R, Varma S, Gray S, Lee S-B (1998) Containment of herbicide resistance through genetic engineering of the chloroplast genome. Nat Biotechnol 16:345–348

De Cosa B, Moar W, Lee S-B, Miller M, Daniell H (2001) Overexpression of the *Bt cry*2Aa2 operon in chloroplasts leads to formation of insecticidal crystals. Nat Biotechnol 19:71–74

Degtyarenko KN, North AC, Findlay JBC (1999) PROMISE: a database of bioinorganic motifs. Nucleic Acids Res 27:233–236

Ding S-W, Rathjen JP, Li W-X, Swanson R, Healy H, Symous RH (1995) Efficient infection from cDNA clones of cucumber mosaic cucumovirus RNAs in a new plasmid vector. J Gen Virol 76:459–464

Dobrzanska M, Krysiak C, Kraszewska E (1997) Transient and stable transformation of wheat with DNA preparations delivered by a biolistic method. Acta Physiol Plant 19:277–284

Fu X, Duc LT, Fontana S, Bong BB, Tinjuangjun P, Sudhakar D, Twyman RM, Christou P, Kohli A (2000) Linear transgene constructs lacking vector backbone sequences generate low-copy-number transgenic plants with simple integration patterns. Transgenic Res 9:11–19

Gorbunova V, Levy AA (1999) How plants make ends: DNA double-strand break repair. Trends Plant Sci 4:263–269

Hanin M, Mengiste T, Bogucki A, Paszkowski J (2000) Elevated levels of intrachromosomal homologous recombination in Arabidopsis overexpressing the MIM gene. Plant J 24:183–189

Haseloff J, Amos B (1995) GFP in plants. Trends Genet 11:328–329

Hensgens LA, de Bakker EP, van Os-Ruygrok EP, Rueb S, van de Mark F, van der Maas HM, van der Veen S, Kooman-Gersmann H, Hart L, Schilperoort RA (1993) Transient and stable expression of *gusA* fusions with rice genes in rice, barley and perennial ryegrass. Plant Mol Biol 22: 1101–1127

Herbert D, Kikkert JR, Smith FD, Reisch BI (1993) Optimization of biolistic transformation of embryogenic grape cell suspension. Plant Cell Rep 12:585–589

Herzog RW, Daniell H, Singh NK, Lemke PA (1996) A comparative study on the transformation of *Aspergillus nidulans* by microprojectile bombardment of conidia and a more conventional procedure using protoplast treated with polyethyleneglycol. Appl Microbiol Biotechnol 45:333–337

Higgins ES (1956) Biochemical studies on molybdenum. PhD Diss, Syracuse University, Syracuse, NY

Higgins ES, Richert DA, Westerfeld WW (1956a) Molybdenum deficiency and tungstate inhibition studies. J Nutr 59:539–559

Higgins ES, Richert DA, Westerfeld WW (1956b) Tungstate antagonism of molybdate in *Aspergillus niger*. Proc Soc Exp Biol Med 92:509–511

Holmes VF, Cozzarelli NR (2000) Closing the ring: links between SMC proteins and chromosome partitioning, condensation, and supercoiling. Proc Natl Acad Sci USA 97:1322–1324

Hu Y, Faham S, Roy R, Adams MWW, Rees DC (1999) Formaldehyde ferrodoxin oxidoreductase from *Pyrococcus furiosus*: the 1.85 A resolution crystal structure and its mechanistic implications. J Mol Biol 286:899–914

Ingram HM, Livesey NL, Power JB, Davey MR (2001) Genetic transformation of wheat: progress during the 1990s into the millennium. Acta Physiol Plant 23:221–239

Ivin KJ, Mol JC (1997) Olefin metathesis and metathesis polymerization. Academic Press, San Diego

Johnson JL, Rajagopalan KV, Cohen HJ (1974) Molecular basis of the biological function of molybdenum. Effect of tungsten on xanthine oxidase and sulfite oxidase in the rat. J Biol Chem 249:859–866

Johnson JL, Rajagopalan KV, Mukund S, Adams MWW (1993) Identification of molybdopterin as the organic component of the tungsten cofactor in four enzymes from hyperthermophilic archaea. J Biol Chem 268:4848–4852

Johnson PH, Grossman LI (1977) Electrophoresis of DNA in agarose gels. Optimizing separations of conformational isomers of double- and single-stranded DNAs. Biochemistry 16:4217–4225

Kemper EL, da Silva MJ, Arruda P (1996) Effect of microprojectile bombardment parameters and osmotic treatment on particle penetration and tissue damage in transiently transformed cultured immature maize (*Zea mays* L) embryos. Plant Sci 121:85–93

Kisker C, Schindelin H, Rees DC (1997) Molybdenum cofactor-containing enzymes: structure and mechanism. Annu Rev Biochem 66:233–267

Klein TM, Wolf ED, Wu R, Sanford JC (1987) High-velocity microprojectiles for delivering nucleic acids into living cells. Nature 327:70–73

Klein TM, Arenzen R, Lewis PA, Fizpatrick-McEligott S (1992) Transformation of microbes, plants and animals by particle bombardment. Biotechnology 10:286–291

Kleter GA, van der Krieken WM, Kok EJ, Bosch D, Jordi W, Glissen LJ (2001) Regulation and exploitation of genetically modified crops. Nat Biotechnol 19:1105–1110

Krysiak C (2002) Genotoksycznosc wolframu w warunkach biolistycznej transformacji roslin. (Genotoxicity of tungsten under conditions of biolistic transformation of plants). Ph.D. dissertation, Instytut Biochemii i Biofizyki PAN, Warszawa, Poland

Krysiak C, Mazus B, Buchowicz J (1999a) Relaxation, linearization and fragmentation of supercoiled circular DNA by tungsten microprojectiles. Transgenic Res 8:303–306

Krysiak C, Mazus B, Buchowicz J (1999b) Generation of DNA double-strand breaks and inhibition of somatic embryogenesis by tungsten microparticles in wheat. Plant Cell Tissue Organ Cult 58:163–170

Kuiper HA, Kleter GA, Noteborn HP, Kok EJ (2001) Assessment of the food safety issues related to genetically modified foods. Plant J 27:503–528

Labra M, Savini C, Bracale M, Pelucchi N, Colombo L, Bardini M, Sala F (2001) Genomic changes in transgenic rice (*Oryza sativa* L.) plants produced by infecting calli with *Agrobacterium tumefaciens*. Plant Cell Rep 20:325–330

Lambe P, Dinant M, Matagne RF (1995) Differential long-term expression and methylation of the hygromycin phosphotransferase (*hph*) and beta-glucuronidase (GUS) genes in transgenic pearl millet (*Pennisetum glaucum*) callus. Plant Sci 108:51–62

Levy RB, Boudart M (1973) Platinum-like behavior of tungsten carbide in surface catalysis. Science 181:547–549

Ljungdahl LG, Andreesen JR (1975) Tungsten, a component of active formate dehydrogenase from *Clostridium thermoaceticum*. FEBS Lett 54:279–282

Maheshwari N, Rajyalakshmi K, Baweja K, Dhir SK, Chowdhry CN, Maheshwari SC (1995) In vitro culture of wheat and genetic transformation – retrospect and prospect. Crit Revs Plant Sci 14: 149–178

Mazus B, Krysiak C, Buchowicz J (2000) Tungsten particle-induced nicking of supercoiled plasmid DNA. Plasmid 44:89–93

Meins F (2000) RNA degradation and models for post-transcriptional gene silencing. Plant Mol Biol 43:261–273

Mendel RR, Schwarz G (1999) Molybdoenzymes and molybdenum cofactor in plants. Crit Rev Plant Sci 18:33–69

Mengiste T, Paszkowski J (1999) Prospects for the precise engineering of plant genomes by homologous recombination. Biol Chem 380:749–758

Meyer P, Saedler H (1996) Homology-dependent gene silencing in plants. Annu Rev Plant Physiol Plant Mol Biol 47:23–48

Miller AC, Mog S, McKinney L, Luo L, Allen J, Xu J, Page N (2001) Neoplastic transformation of human osteoblast cells to the tumorigenic phenotype by heavy metal-tungsten alloy particles: induction of genotoxic effects. Carcinogenesis 22:115–125

Mittelsten Scheid O, Paszkowski J (2000) Transcriptional gene silencing mutants. Plant Mol Biol 43:235–241

Mittler R, Rizhsky L (2000) Transgene-induced lesion mimic. Plant Mol Biol 44:335–344

Moellenbeck DJ, Peters ML, Bing JW, Rouse JR, Higgins LS, Sims L, Nevshemal T, Marshall L, Ellis RT, Bystrak PG, Lang BA, Stewart JL, Kouba K, Sondag V, Gustafson V, Nour K, Xu D,

Swenson J, Zhang J, Czapla T, Schwab G, Jayne S, Stockhoff BA, Navra K, Schnepf HE, Stelman SJ, Poutre C, Koziel M, Duck N (2001) Insecticidal proteins from *Bacillus thuringiensis* protect corn from corn rootworms. Nat Biotechnol 19:668–672

Mukund S, Adams MWW (1991) The novel tungsten-iron-sulfur protein of the hyperthermophilic archaebacterium, *Pyrococcus furiosus*, is an aldehyde ferrodoxin oxidoreductase. J Biol Chem 266:14208–14215

Mukund S, Adams MWW (1993) Characterization of a novel tungsten-containing formaldehyde ferrodoxin oxidoreductase from the hyperthermophilic archaeon, *Thermococcus litoralis*. J Biol Chem 268:13592–13600

Murry LE, Elliott LG, Capitant SA, West JA, Hanson KK, Scarafia L, Johnston S, DeLuca-Flaherty C, Nichols S, Cunanan D, Dietrich PS, Mettler IJ, Dewald S, Warnick DA, Rhodes C, Sinibaldi RM, Brunke KJ (1993) Transgenic corn plant expressing MDMV strain B coat protein are resistant to mixed infections of maize dwarf mosaic virus and maize chlorotic mottle virus. Biotechnology 11:1559–1564

Muskens MWM, Vissers APA, Mol JNM, Kooter JM (2000) Role of inverted DNA repeats in transcriptional and post-transcriptional gene silencing. Plant Mol Biol 43:243–260

Notton BA, Hewitt EJ (1971) The role of tungsten in the inhibition of nitrate reductase activity in spinach (*Spinacea oleracea* L.) leaves. Biochim Biophys Res Commun 44:702–710

OhHalloran TV (1993) Transition metals in control of gene expression. Science 261:715–725

Palauqui J-P, Balzergue P (1999) Activation of systemic acquired silencing by localized introduction of DNA. Curr Biol 9:59–66

Patnaik D, Khurana P (2001) Wheat biotechnology: a minireview. Electronic J Biotechnol 4, online at http://www.ejb.org/cntent/vol4/issue2/full4/

Pawlowski WP, Somers DA (1998) Transgenic DNA integrated into the oat genome is frequently interspersed by host DNA. Proc Natl Acad Sci USA 95:12106–12110

Paximadis M, Idris AM, Torres-Jerez I, Villarreal A, Rey MEC, Brown JK (1999) Characterization of tobacco geminiviruses in the Old and New World. Arch Virol 144:703–717

Puchta H, Hohn B (1996) From centiMorgans to base pairs: homologous recombination in plants. Trends Plant Sci 1:340–348

Register JC III, Peterson DJ, Bell PJ, Bullock WP, Evans IJ, Frame B, Greenland AJ, Higgs NS, Jepson I, Jiao S, Lewnau CJ, Sillick JM, Wilson HM (1994) Structure and function of selectable and non-selectable transgenes in maize after introduction by particle bombardment. Plant Mol Biol 25:951–961

Rich T, Allen RL, Wyllie AH (2000) Defying death after DNA damage. Nature 407:777–783

Russell JA, Roy MK, Sanford JC (1992) Physical trauma and tungsten toxicity reduce the efficiency of biolistic transformation. Plant Physiol 98:1050–1056

Ryerson DE, Heath MC (1996) Cleavage of nuclear DNA into oligonucleosomal fragments during cell death induced by fungal infection or by abiotic treatments. Plant Cell 8:393–402

Sambrook J, Fritsch EF, Maniatis T (1989) Molecular cloning: a laboratory manual. Cold Spring Harbor Laboratory Press, Cold Spring Harbor, New York

Sanford JC (1988) The biolistic process. Trends Biotechnol 6:299–302

Sanford JC, Smith FD, Russell JA (1993) Optimizing the biolistic process for different biological applications. Methods Enzymol 217:483–509

Schaverien CJ, Dewan JC, Schrock RR (1986) A well-characterized, highly active, Lewis acid free olefin metathesis catalyst. J Am Chem Soc 108:2771–2773

Schmitz RA, Albracht SPJ, Thauer RK (1992) Properties of the tungsten-substituted molybdenum formylmethanofuran dehydrogenase from *Methanobacterium wolfei*. FEBS Lett 309:78–81

Schweizer P, Pokorny J, Schulze-Lefert P, Dudler R (2000) Double-stranded RNA interferes with gene function at the single-cell level in cereals. Plant J 24:895–903

Smith FD, Harpending PR, Sanford JC (1992) Biolistic transformation of prokaryotes: factors that affect biolistic transformation of very small cells. J Gen Microbiol 138:239–248

Tyagi AK, Mohanty A (2000) Rice transformation for crop improvement and functional genomics. Plant Sci 158:1–18

Van der Fits L, Memelink J (1997) Comparison of the activities of CaMV 35S and FMV 34S promoter derivatives in *Catharanthus rescus* cells transiently and stably transformed by particle bombardment. Plant Mol Biol 33:943–946

Van der Maas HM, de Jong ER, Rueb S, Hensgens LAM, Krens FA (1994) Stable transformation and long-term expression of the gusA reporter gene in callus lines of perennial ryegrass (*Lolium perenne* L.). Plant Mol Biol 24:401–405

Vasil IK (1994) Molecular improvement of cereals. Plant Mol Biol 25:925–937

Vasil V, Srivastava V, Castillo AM, Fromm ME, Vasil IK (1993) Rapid production of transgenic wheat plants by direct bombardment of cultured immature embryos. Biotechnology 11: 1548–1558

Wang H, Li J, Bostock RM, Gilchrist DG (1996) Apoptosis: a functional paradigm for programmed plant cell death induced by a host-selective phytotoxin and invoked during development. Plant Cell 8:375–391

Wengrovius JH, Schrock RR, Churchill MR, Missert JR, Youngs WJ (1980) Tungsten-oxo alkylidene complexes as olefin metathesis catalysts and the crystal structure of $W(O)(CHCMe_3)(PEt_3)Cl_2$. J Am Chem Soc 102:4515–4516

Wetterauer B, Salger K, Demel P, Koop H-U (2000) Efficient transformation of *Dictyostelium discoideum* with a particle inflow gun. Biochim Biophys Acta 1499:139–143

Wide M, Danielsson BR, Dencker L (1986) Distribution of tungstate in pregnant mice and effects on embryogenic cells in vitro. Environ Res 40:487–498

Woese CR, Kandler O, Wheelis ML (1990) Towards a natural system of organisms: proposal for the domains Archaea, Bacteria, and Eucarya. Proc Natl Acad Sci USA 87:4576–4579

Xu D, Duan X, Wang B, Hong B, Ho TD, Wu R (1996) Expression of a late embryogenesis abundant protein gene, *HVA1*, from barley confers tolerance to water deficit and salt stress in transgenic rice. Plant Physiol 110:249–257

Yamamoto I, Saiki T, Liu SM, Ljungdahl LG (1983) Purification and properties of NADP-dependent formate dehydrogenase from *Clostridium thermoaceticum*, a tungsten-selenium-iron protein. J Biol Chem 258:1826–1832

Yao N, Tada Y, Park P, Nakayashiki H, Tosa Y, Mayama S (2001) Novel evidence for apoptotic cell response and differential signals in chromatin condensation and DNA cleavage in victorin-treated oats. Plant J 28:13–26

Yasuda H, Nakajima M, Ito T, Ohwada T, Masuda H (2001) Partial characterization of genes whose transcripts accumulate preferentially in cell clusters at the earliest stage of carrot somatic embryogenesis. Plant Mol Biol 45:705–712

Yepes LM, Mittak V, Pang S, Gonsalves C, Slightom JL, Gonsalves D (1995) Biolistic transformation of chrysanthemum with the nucleoprotein gene of tomato spotted wilt virus. Plant Cell Rep 14:694–698

Zimmerman JL (1993) Somatic embryogenesis: a model for early development in higher plants. Plant Cell 5:1411–1423

Zoubenko OV, Allison LA, Svab Z, Maliga P (1994) Efficient targeting of foreign genes into the tobacco plastid genome. Nucleic Acids Res 22:3819–3824

Subject Index

A
Accelerator (gun) 128–129
ACCELL (technology, particles from the electric discharge particle device) 160, 164
Acetosyringone 30, 56
Agrobacterium rhizogenes 23–39
Agrobacterium suspension drop method 45–49
Agrobacterium-mediated transformation 76, 94–96
- cereal plants 70
- requirements 45
Agrolistics 167
Agropine 24, 26
Alfalfa, root hair growth 28
Alginate, embedding protoplasts 75
Alkaloid production 27
Alliinase expression, silence 55
Alliinase promotor-driven anstisense bulb alliinase 62
Allium 53, 56
Allium transformation 56
Alternaria helianthi 110
Amplifying DNA, standard PCR conditions 156
Antibiotic resistence gene sequences 149
Antibiotic resistence markers 32
Antibiotic selection, transformation 56–57
Antirrhinum majus 36
Antisense alliinase gene epression 61–62
Apical dominance 36
Apoptosis 185
Arabidopsis thalliana 26–28
Arbitrary degenerated primer 60
Aspergillus niger 176
Autochthonous forests 141

B
Bacillus thuringiensis 179
Backbone-free PTU-containing DNA fragments 157
Bacteria, use to bioremediate polluted soils 5
Bacterial inoculum, preparation 157
Bar gene expression, stability 132, 133
Barley explants, electroporation 74–82
Barley-wheat hybrids 82
Basta medium 48
Basta resistance phenotype, transmission 48
Bialophos 93
Binary plasmid 47
Binary T-DNA 31–33
Binary vector 56, 58
Bioinformatics 5
Biolistic conditions, optimization 186
Biolistic inoculation 181–182
Biolistic transformation 175–188
- tungsten as DNA carrier 178–181
Biolistic treatments 187
Biolistics 53, 76, 159–170
- advantage 166
Biological matters, rights to 20–21
Biotechnological innovation 2–6
Biotechnology, intellectual property rights 1–21
Bombarded pollen
- dehydration 136
- long-term storage 136
- used in pollination 134, 135
Bombardment, age of culture 162
Brassica napus 159

C
Callus proliferation medium (CPM) 64
Calluses, browning during selection 64
Carbenicillin 30, 37
Carrier DNA, addition to electroporation buffer 73
Cefotaxime 30
Cell fusion-mediated transformation 53
Cell totipotency, effect of particle bombardment 186
Cell viability, determination by electrical field strength 73

Cellular DNA
- double-strand breaks (DSBs) 175
- fragmentation, detection 184, 185
Challenges of various property regimes 17–21
Chromosomal instability, genotype-dependent 161, 166
Chromosome analysis, regnerated transgenic plants 122
Chromosome decondensation 113
Chromosome uptake 111
Circular plasmid DNA, addition to electroporation 79
Circular plasmids, circulation, by tungsten 187
Clavulanic acid 30
Clostridium thermoaceticum 176, 177
Coating protocol 181
Co-bombardment 100
Co-bombardment plasmids 167
Co-cultivated embryos 58
Co-cultivation 29, 30, 64
- of pollen with *A. tumefaciens* 46
- step in *Agrobacterium*-mediated transformation 95
- with *Agrobacterium* 53–65
Consequential gene silencing 132
Convention of Biological Diversity (1992) 20
Conventional breeding, combination with transgenic breeding 110
Copyright 11
- protection 14–16
Co-transformation 31–33, 100
- binary T-DNA 31–33
- system 101, 102
Co-transformed events 101, 102
Criteria, substantive, patents 11–13
Cryoprotection solution 150
Cucumopine 24, 26
Culture periods, electroporation 80
Cytokinin 23, 35
- levels 30

D
Data base protection 14–16
Defense genes 137–139
Defensin 138
Defensin-like protein, encoding 140
Desiccation, enhancement germination embryos 163, 164
Detectable transgene expression 73
Dilution bacterial suspension 29
Direct gene transfer, by electroporation 71–74
Directive on the legal protection of biotechnological inventions 13
Disclosure, procedural criterion 13
Diversity, maintenance 141
DNA analysis, Southern blot 49
DNA constructs, integration 179, 180
DNA delivery, biolistic 159
DNA delivery methods 93
DNA delivery via whiskers 152–153
DNA-carrying ability 164
DNA-coated tungsten particles, transformation 179
Double nicking (linearization DNA strand) 184
Double-stranded DNA (dsDNA), fluorescence monitor 154
Dysfunction somatic embryogenesis 160

E
Early embryonic stage 150
Economic reality, patent system 19–20
Ectopic expression 139
Electrical variables, electroporation 73–74
Electroporation 69–84
- intact tissue 77–82
- timing 71
Electroporation buffer (EB) 72
Electroporation parameters, optimization 76
Electroporator, commercial 77, 79
Elimination markers genes, strategy 101
Elimination markers, methods 100–101
Elimination of marker genes from transgenic plants 99–102
Elongation medium (EM) 39
Elution buffer, FPLC 148
Embedding Whisker-treated cells 153
Embryogenic capacity, electroporated tissues 72
Embryogenic cultures 128–133
Embryogenic suspension culture 74, 150–152
- from callus 150–152
Embryonic cultures, protocol 168
Enzymatic treatment, before electroporation 72
Enzymic assay gene transfer sunflower protoplasts 119–120
Erwinia 55, 139
Escape nontransgenic regenerants 61
Escherichia coli 148, 179
Estimation of transgene copy number 153

Ethylene inhibitors, influence on regeneration 35
Eucalyptus 137
Exclusive rights 6, 7
Explant orientation 161
Explants, choice 28
Expressed sequence Tags (ESTs) 3
Expression of *rol* genes 31
Extending culture generation time, prolongation youth state 161

F
Fairness to providers of biological matter 20, 21
Farmers privilege 16
Fast protein liquid chromatography (FPLC) 148
Fertile transgenic barley plants 70
Fertile transgenic wheat 83
FISH (fluorescence in situ hybridization) 185
Floral dip method 46
Flow cytometric analysis, genome size 120–122
Fluorescence *in situ* hybridization (FISH) 185
Foreign gene insertion 99
Forest ecosystems 141
FPLC (fast protein liquid chromatography) 148

G
Garlic transformation 53–65
- protocol 63, 64
Gene delivery, by acceleration of DNA-coated particles 164
Gene expression in T_O plants, foreign 98, 99
Gene flow, ecological impact 142
Gene integration, effect of bombardment of different quantities DNA 167
Gene silencing 99
Gene transfer process 110
Gene transfer
- natural 142
- PEG-mediated 109–124
- to *Petunia*, pollen on receptive stigma 49
Generating hairy roots, efficiency 36
Genes modifying susceptibility 137
Genes regulation embryogenesis 136, 137
Genetic analysis of the progeny 99
Genetic information, cells 3, 4
Genetic transformation, by biolistics 159–170
Geneticin 80, 82, 129
Genome 5
Genomic DNA extraction, maize callus 154
Genomic DNA from transgenic plants, Southern blot 97, 98
Genotoxicity, tungsten 175
Genotoxins 175
Genotypic specificity, regeneration 160
Germination mature embryo 163, 164
Germline transformation 46, 47
Gfp (green-fluorescent protein) 32
- gene 57
- markers 32
- selection 166
β-glucuronidase assay 48, 49
β-glucoronidase (*gus*) 32
- enzyme assay 119, 120
Glycine max 159
Glyphosate 54
- resistance gene 61
- tolerant plants, generation 58
Gold particle, coated 129, 130, 137
Green calli, generation 27
Green-fluorescent protein (gfp, GFP) 32, 69
Gun (accelerator) 128
GUS assay buffer 120
GUS casette 137
GUS expression in pollen 135
GUS gene expression 80–82
GUS transient expression, in sorghum embryo 96

H
Hairy root lines, propagation in liquid cultures 33, 34, 39
Hairy root syndrome 23
Hairy roots
- chimeric nature 35
- clonal status 33
- transformed nature 31
Helianthus annuus 109–124
Herbicide resistence
- expression 61
- onion germplasm 54
- *Petunia* seedlings, screening 47, 48
Heterobasidion annosum 127, 137, 138
Histodifferentiation embryo 163
Homebox gene 137
Hordeum vulgare 69–84
Hormone balance 34
- formation of transformed roots 23
Human Genome Organization Ethics 20
Human Genome Project, protection 14–16
Hygromycin 165
Hypervirulent *A. tumefaciens* strains 30

I
Immaterial aspects of inventions 4
Immaterial information 2
Immature embryos, isolation 56
In situ hybridization (ISH), distribution of gene sequences 122, 123
Incubation wounded explants 29
Industrial application, criterion patents 12, 13
Infection impact T-DNA 95
Infection
- of explants 29
- steps in manipulation of *Agrobacterium*-mediated transformation 95
Insect resistance, effective transgenic approaches 55
Intact tissue electroporation 74
Integration of *rol* genes 31
Integration of T-DNA vector backbone sequence into plant genome 98
Integration of transgenes 94
Intellectual property laws, application to biological material 1
Intellectual property rights (IPRs) 6–14
Interaction among forest species 141
Invention 9, 10
Inventive step, criterion patents 12
Inventor 11
Isolated DNA, shuttle into host cells 110
Isolation of nuclei 121
Isopentenyltransferase (ipt) 36

K
Kanamycin 32

L
Large DNA molecule uptake 112–114
Leek transformation 53–65
Lignin 139
Lipid transfer protein 136
Liquid pollination 135
Long-term hairy root cultures, stability 33
Low copy DNA integration 147
Luc gene coding for luciferase 128
Luciferase 32
Lysine content 91

M
Macroprojectile 128
Magainin 55
Magainin-expressing potato plants 55
Maize ubiquitine promotor 137, 138
Manipulation of sorghum transformation, factors 95
Mannitol pre-treatment of donor anthers 77
Mannopine 24, 26
Marker-free transgenics, in *Sorghum*, importance 100
Master-Mixture, PCR, genomic DNA 154, 155
Maturation embryo 163
- use for electroporation 80
Maturation medium, embryo soyebean 170
MAT-vector 101
- system 37
Medicago, truncatula 46
Meiotic analysis flower buds transgenic plants 122
Methods to eliminate markers from transgenic plants 100, 101
*m-gfp*ER marker 57, 60, 61
Microcally, sunflower 114
Microfusion solution 112
Microinjection, haploid cells 76
Micronucleated protoplast 113
Micronuclei uptake 111
Microprojectile bombardment 127–142
- transformation 93, 94
Microprotoplasts 112, 113
Microspore cultures, gene transfer 76
Mikimopine-type plasmid 26
Molecular analysis, transgenic plants 115
Multi-auto transformation (MAT) vector systems 35–36
Multiple copies of integrated DNA 153
Multi-autotransformation (MAT), *rol* type 25
Mustard, root hair growth 28
Mutations, gene transfer experiments 115

N
Negative markers 33
Nested primers 59, 60
Nicking (apparent relaxation) 184
Nicking-closing (relaxation) DNA strand 184
Nicotiana langsdorffii 46
Non-oncogenic *A. tumefaciens* 30
Novelty, criterion patents 12
Nucleases, microspores 134
Nurse cell layers, plant derived 30

O
Oilseed rape, root hair growth 28
Onion transformation 53–65
Opening of hydrophilic pores, membrane, to electric pulses 71
Opine biosynthesis 24, 26

Opines 31
Osmotic balance 73
Osmotic pretreatment, Whiskers transformation 152
Osmoticum
- exposing before bombardment 129
- for tissue electroporation 71, 72

P

Paramomycin 32
Particle accelerator, types 127, 128
Particle bombardment 127–142
- mediated gene transfer 110
- soybean tissues 160
- transformation pollen 46
Particle inflow gun 128, 130
Patent Cooperation Treaty 9
Patentability, requirements 11
Patents, biotechnology 8–14
Pathogenesis-related (PR) proteins 138
PCR, thermal asymmetric interlaced (TAIL-PCR) 59–61
PEG-mediated gene transfer 111
Petunia 45–49
phenotypic alteration 54
Phloroglucinol 139
Phomopsis helianthi 110
Phosphinothricin, contact herbicide 54, 58, 93
Physical embodiment of gene, protection 2
Physical innovations 3, 4
Phytohormones, regeneration process 73
Phytoremediation, root-mediated 25
Picea abies 127–142
Pinus radiata 135, 141
Pinus sylvestris 134
Plant fertility, effect of bombardment 167
Plant regeneration *Arabidopsia* 27
Plant regeneration from hairy roots, protocol 38, 39
Plant transcription unit (PTU) 98, 147–149
Plant variety protection 16, 17
Plasmid DNA, digestion 148
Plasmid
- *bar* gene 137
- uptake 111
Plasmopara halstedii 110
Pollen chamber, spruce 134
Pollen flow 142
Pollen genotypes, spectrum 135
Pollen germination medium 47, 123
Pollen transformation 133–136
Polyethylene glycol (PEG) technique, gene transfer 109–124
Polymerase chain reaction (PCR), control of DNA transfer 117
Post-bombardment DNA nicking 184
Post-bombardment inhibition 186, 187
Post-electroporation period 71–73
Pre-electroporation period 71–73
Presence of transgenes, molecular proof 99
Primary transformant 62
- transfer in glasshouse 58
Processes using biological matters, grant property 5
Production of chimeras 115
Production of transformed roots, media 38
Proembryogenic masses 137
Progeny of transgenic plants, analysis 96–99, 157
Progeny, molecular analysis 99
Proliferation medium, embryonic suspension cultures 129, 130
Proliferative phase, arrest 161, 162
Promotor activity assays 128
Promotor trapping strategies 25
Promotor-*gusA* construct 128
Propert rights, complex higher life forms 4
Property regimes, biological sciences 21
Property rights
- justification 18
- understanding 21
Protection industrial property 9
Proteomes 5
Protoplast electroporation 74
Protoplast suspension 75
Protoplast systems, electroporation 71, 74–76
PTU 98, 147–149
- containing fragments 148, 149
- fragment purification 149
Puccinia helianthi 110
Purified DNA fragments
- preparation 147–149
- restriction analysis 149
Pyrococcus furiosus 176, 177
Pyiium dthmorphum 137

R

Radish, root hair growth 28
Random amplified polymorphic DNA (RAPD-PCR), fingerprinting DNA tansfer 117, 118
Reaction mixture, PCR, DNA molecule transfer sunflower 117
Recalcitrance soybean transformaton 167
Recombination systems, site-specific 100
Recombination/recognition site (R/RS) 36

Regeneration 34, 35
- of transgenic plants 111
Regeneration process 114
- phytohormones 73
Regeneration soybean embryos, protocol 170
Remedies, to inventor 14
Repetitive proliferation somatic embryogenic culture 161
Reporter genes, transient expression 134
Reporter markers 73
Reproductive biology, spruce 133, 134
Ri plasmid 31–33
Ri T-DNA-associated gene expression 34
Right to things 6, 7
Rights granted to the creator 18
Risk assessment genetic modification 141, 142
RNA interference constructs (RNAi) 54
Rol gene expression 25, 35
Root hair production, *Arabidopsis* 27, 28
Root-inducing medium 35
Root-inducing plasmid (*Ri*) 23, 24
Rooting ability, increase 34
Routine genetic transformation technology 91

S
Saccharomyces cerevisiae 100, 179
Sclerotial germination, inhibition 55
Sclerotinia 110, 123
Sclerotium cerpivorum 54
Segregation of transgenes in subsequent generations 101
Selectable genes 56
Selectable marker genes, elimination from transgenic plants 99
Selectable marker, *bar* gene 93
Selectable resistance gene 165
Selection soybean tissue, protocol 169
Self fertilization of transgenic plants by greenbottle flies 59
Semi-solid differentiation medium 114
Shoot formation, spontaneous 35
Shoot regeneration 151
Shoot-inducing medium (SIM) 38
- solified 64
Short DNA molecules uptake 111, 112
Silencing genes 94
Silencing of transgene, post-transcriptional 49
Silencing response 62
Single-strand breaks (SSBs) 175
Single-stranded (ss) copy of the T-DNA 23
Somatic embryogenesis 136
- effect of particle bombardment 187
Somatic hybrid regeneration 110
Sonication-assisted *Agrobacterium*-mediated DNA delivery 159
Sorghum transformation 91–102
Southern assay in T_0plants 96
Southern blot analysis, transgenic sunflower calli 116
Sox-9 gene 2
Soybean, genetic transformation 159–170
Specific amplification products, quantification 153
Spodoptera exigua 55
Stability
- long-term hairy root cultures 34
- transgene expression 60–62
Stable monopoly, patent 19
Stable transformation higher plants with DNA-coated tungsten particles 180, 181
Stable transformation, requirements 96
Stable transformed transgenic plants 129–131
Standard curve transgene copy number, by qRT-PCR 154, 155
Sterility, plant regeneration 160
Subculture interval, regeneration 95
Sugar beet, root hair growth 28
Super-binary vector 94, 101
Suppressed virulence 29
Suppression of reproduction 142
Surface sterilization seeds 47, 48, 56
Suspension culture
- cryopreservation 150
- medium 150
Suspensor, expression of aquaglyceroporin 137
Systemic herbicide glyphosate 54

T
TAIL-PCR 59–61
Target tissue optimization 165
Target tissue transformation 95
T-DNA (transfer DNA) 23, 24, 26, 31–33, 35–37, 45
2 T-DNA 101
Ternary plasmid 37
Timentin 30, 57, 64
Tissue electroporation, experimental steps 78
T_0 plants, molecular analysis 97, 98
Tobacco, root hair growth 28
Tomato, root hair growth 28
Toxic radicals, production 138

Trade Related Aspects of Intellectual Property Rights (TRIPs) 9
Trade secrecy 18, 19
- protection 8, 11, 13
Trade secrets 7, 8
Transfer DNA fragments, Southern blot analysis 116
Transferring polygenic determined traits 113
Transformability of a line 162
Transformants, analysis 47–49
Transformation
- *Agrobacterium rhizogenes* 37
- *Agrobacterium*-mediated 23–39
- by co-cultivation 53–65
- DNA-coated tungsten particles 180, 181
- embryogenic cultures 128–133
- of leaf discs 159
- *Sorghum bicolor* 91–102
- soybean 164–167
- by stigma treatment 47
- use of tungsten particles 179
Transformation efficiency 124, 160
- *A. rhizogenes* 26–30
Transformation frequencies, temperatur dependence 30
Transformation maize, Whiskers-mediated 147–157
Transformation *Petunia* 45–49
Transformation pollen 133–136
Transformation procedure
- *Agrobacterium* 56, 57
- garlic 63, 64
Transformation protocols, onion 55–62
Transformation soybean, protocol 168, 169
Transformation technology, *Allium* 53–55
Transformation using herbicide selection 57, 58
Transformed hairy roots, production, protocol 37, 38
Transformed wild-type shoots 36
Transgene anomalies, overcome 166
Transgene copy number estimation 153–156
- Light-Cycler readout, qRT-PCR 156
Transgene detection 59, 60
Transgene expression, stability 131, 132, 135
Transgene fragments, demonstration of presence 59
Transgene integration 187
Transgenes 94, 98
- integration 135
- various categories 142
Transgenic breeding 109, 110
Transgenic callus culture, medium 157
Transgenic clones 165
Transgenic lines, fingerprinting 59
Transgenic nodules 25
Transgenic onions 54, 57
Transgenic plant
- cytogenetic analysis 120–123
- detection, isoenzyme analysis 119
- marker-free 99–102
- morphological analysis 123
- multiple molecular forms of enzymes 118, 119
- production, spruce 131, 132
- progeny, analysis 96–99
- recovery 166, 167
- regeneration 157
- sunflower, DNA extraction 116, 118, 121
Transgenic progeny 157
Transgenic shade house 58
Transgenic shoots, chimeric 36
Transgenic spruce plants 131
Transgenic sublines 138, 139
Transgenic sunflower 109–124
Transient expression 128, 129
- mitotic index peak 165
- in pollen 134, 135
- promotor activity 128, 129
- spruce 129
Transient GFP expression 63
Transient *gus* expression 34
Transposable elements 100
Trans-zeatin 30
Tritordeum plants of barley 82
Tumor-inducing plasmid (Ti) 23
TUNEL (terminal deoxynucleotidyl transferase-mediated dUTP nick end-labelling) 184, 185
Tungsten carbide, catalyst 176
Tungsten compounds, catalytic activity 176
Tungsten cytotoxicity 179
Tungsten genotoxicity 188
Tungsten microparticles, applications 178–182
Tungsten microparticles, genotoxic effects 175–188
Tungsten, biological significance 175–178
Tungsten-damaged plasmid DNA, electrophoretic analysis 182–184
Tungsten-DNA interaction 177, 178
Tungsten-induced changes in electrophoretic mobility 182
Tungsten-induced DNA damage 178
Tungsten-induced DNA lesions 182–185
Tungsten-promoted strand scission 187, 188
Tungstoenzymes 176, 177

U
Ubiquitin promotors 129, 131, 137
United States Patent and Trademark Office (USPTO) 12
Unpatentable processes 10
Utility, criterion patents 12

V
Vector design 101
Vector of genetic transformation, pollen 133
Vir gene products 31
Viral cDNA-derived construct, tungsten particle coated 181
Virulence gene-inducing factor 56
Virulence proteins 45
Visual reporter genes 60, 61
Visual selection, transformation 56, 57
Volume-driven thinking, forestry 140

W
Whisker
- dry silicon carbide 147–157
- supplyer 152
- transformation 152
- treated cells, recovery 153
- treatment, delivers DNA to surface cell layers 150
Wounded plants, inoculation 26

Y
Yield of transformed cells 113

Z
Zea mays 49
Zygosaccharomyes 36

Printing: Mercedes-Druck, Berlin
Binding: Stein+Lehmann, Berlin